SCIENCE,

TECHNOLOGY,

and

FREEDOM

SCIENCE,

TECHNOLOGY,

and

FREEDOM

edited by

Willis H. Truitt

T. W. Graham Solomons

University of South Florida

with the collaboration of
Anne J. Truitt

HOUGHTON MIFFLIN COMPANY BOSTON

Atlanta Dallas Geneva, Illinois Hopewell, New Jersey Palo Alto London

Art Credits

Page xiv, part of a drawing by John Clift.
Page 60, *Généalogie* by Frits Van den Berghe; the
Emanuel Hoffman-Stiftung, Kunstmuseum Basel,
Switzerland.
Page 128, *Portrait of Uhde* by Pablo Picasso; from the
collection of Mr. and Mrs. Joseph Pulitzer, Jr.
Page 226, part of the painting *E Pluribus Unum* by
Mark Tobey; Seattle Art Museum.

Printed in the U.S.A.

Library of Congress Catalog Card Number: 73–11766
ISBN: 0–395–17685–9

For Benjamin and Jennie

Contents

4 Science, Technology, and the Environment

Introduction

Science and technology have had an enormous impact on modern society and culture. This book raises some acute questions about this impact. It also provides a number of answers to these questions from varying points of view. The final decisions that are made about the uses of science and technology will undoubtedly shape the future. And this is why we believe that the material presented in this book is worth learning.

For generations it was believed that science and technology would provide the solutions to the problems of human suffering, disease, famine, war, and poverty. But today these problems remain; in fact, many argue that they are expanding. Some even conclude that science and technology, as presently constituted, are not capable of meeting the collective needs of mankind. A more radical position is that modern scientific methods and institutions, because of their very nature and structure, thwart basic human needs and emotions; the catastrophies of today's world, and the greatest threat to its future, are the direct consequence of science and technology.

If this is so, or if only partially accurate, then answers must be found to correct and improve the situation. The importance of science and technology need not be implied, for it is direct, immediate, and basic to everyday life. The general thesis of the book, then, is to explore the pros and cons of the expansion of scientific knowledge and technology into human affairs. The central issues are: Does science presuppose certain values with organizational and political consequences inimical to human freedom and dignity? Or, alternatively: Are science and its technological applications essentially neutral to human aspirations; is science free of value assumptions? Put another way: Should morality be a factor in deciding basically technical issues? If so, what are the legitimate uses of science and technology? How are science and technology being abused, and for what ends, for whose benefit? For the most part, these questions admit no easy answers. Although we are generally persuaded to divorce facts from values, the separation of the two is not always self-evident.

The first chapter of the book presents the implications of the rise of science and technology to the dominant position it commands in contemporary society. How did scientific perspective come to dominate the modern world? Are science and technology inevitably dehumanizing? Are there two cultures, a scientific culture and a nonscientific culture, which are opposed to each other in basic values? These are but some of the issues considered.

The second chapter treats the historical and social origins of science and technology. In tracing their development from early origins, it becomes obvious that science and technology are not autonomous. Science has thrived and expanded rapidly in some societies and cultures, and it has been suppressed and inhibited in others.

The third chapter returns to the contemporary by extending the material presented in Chapter 1 to a consideration of more particular and concrete issues. Special emphasis is given to (1) the role of ethics in scientific practice, (2) the alienation of scientists themselves, (3) the dehumanization caused by the application of technology to work, education, and leisure, (4) the growing chemical and biological manipulation of men and their needs, (5) the problem of the obsolescence of work and the meaning of leisure in the affluent, post-industrial society, and (6) the problems related to enormous population growth made possible by advances in science and technology.

The concluding chapter of the book considers the factors—technical, political, and economic—leading to the current environmental crisis. Here it is asked: Can science and technology be "liberated" from the exploitative role imposed on them by certain narrowly defined interests? Furthermore, questions about the relation of affluence and ever-increasing production to waste and environmental destruction are discussed in this chapter. The effect of modern warfare on the environment is considered in some detail. In general, these selections show the role played by science and technology in environmental pollution as well as the social institutions and practices that encourage destructive applications. Can science and technology alone reverse the pollution and destruction of nature, or must wider solutions be sought?

Many books, including this one, have apparatus to form "bridges" from one chapter or selection to another. However, the most important apparatus needed for the study of this material is an inquiring and critical mind. The rewards will be ample, not just in terms of the immediate satisfaction of grasping some of the most important issues of our time, but also in the long-term application of your knowledge, particularly when you consider what is at stake.

In answer to the point put forth in the following prefatory article by Nicholas Robinson, we have carefully selected scientific readings that we consider to be of both interest and educational value to the nonscience major who is interested in the development of technology as it applies to the human environment.

Thanks are due to the various publishers and authors for permission to reprint the articles we have chosen, and to Robert L. Mortimer for research and assistance in preparing the manuscript.

W.H.T.
T.W.G.S.

NICHOLAS ROBINSON

An Ethic for the Scientific Community

I am not a scientist. I am a law student. And like most people, I do not understand the highly sophisticated and technical science of today. Yet daily I must
rely on science in innumerable ways in this technological society. I do not
believe, however, that the scientific community in this country has shared its
knowledge or assessed the social impact of the mysteries of scientific research
with the people.

I think of the nuclear scientists who, when they developed the awesome power
of the atom and began to understand how to unleash it, were aware that they
had done something awesome, and they shouldered the burden of the moral,
political, and social issues themselves. And they had big worries over these
burdens. The people have been excluded from the decisions on the use of nuclear
power in this country. We do not know Operation Plowshare is a good thing. We
do not know, although I suggest we do know, what the role of the nuclear bomb
is going to be in our time. We have been excluded from the decision-making
right to the ABM vote, and the ABM vote was certainly a confirmation of our
exclusion. As nonscientists, we should not allow it to happen. Scientists must
not preempt the normative and the philosophical and political role in our
society.

The scientific community has alerted us to the problems of the environment
and it has done that increasingly, and increasingly rapidly. This is magnificent.
But scientists must open up all the way. A true scientific community must go
beyond relating science to science, and its members' discoveries to each other.
Scientists must also speak to people like you and me.

The scientific community must frame the context of our education. Scientists
must translate. They must tell us what we need to know; and I suggest that this
is a lifelong education process. And it is an out-of-school process. Moreover, the
education has to reach all of us. The black community is not going to care about
involvement until the blacks understand what they must care about.

Scientists must further frame the policy questions and marshal and present
facts on both sides. Citizens—locally, nationally, and globally—must be in a position to evaluate what is invented. On any given scientific issue, even scientists
perhaps cannot prove an argument to the satisfaction of their opponents. Nevertheless, we need to be exposed to this very debate in terms which we can understand. We need to come to our own conclusions on the basis of sound information.

I believe a critical concern to the scientific community is the government black-

Nicholas Robinson, "An Ethic for the Scientific Community," from *No Deposit-No Return*,
edited by Huey D. Johnson, 1970, Addison-Wesley, Reading, Mass. Pp. 261–262.

listing of certain scientists from participation in commissions. By whose right, in a political way, is nonpolitical information denied to us as citizens? Let the information be heard, and then let the powers that be fight out the philosophies of political decisions. But let this not happen by default or exclusion.

Moreover, I think scientists have an obligation to testify for free in environmental litigation where these suits have been brought. An average environmental suit costs around $100,000 to $150,000 with volunteer help. It is perhaps the most effective way to stop environmental exploitation today. Many scientists, especially in oil and nuclear interests, are under retainer to private interests, and that clearly may conflict with the public interest. I suggest that the scientific community should decide whether its professional obligation lies with the dollar or the environment.

More important, I suggest that scientists develop a new ethic in this crisis. They must march into the committee room, into the conference, into Congress, with scientific respectability. They need an ethic which says that what they do with their science in helping society make decisions is of eminent importance if we are to remain a democracy.

If we are going to have an alert public educated by the scientists, we may survive and make value judgments with our eye toward survival. At present there is no design to our dilemma. We are but flotsam and jetsam destined like the wolf seemingly for extinction.

Science, Technology, and Culture

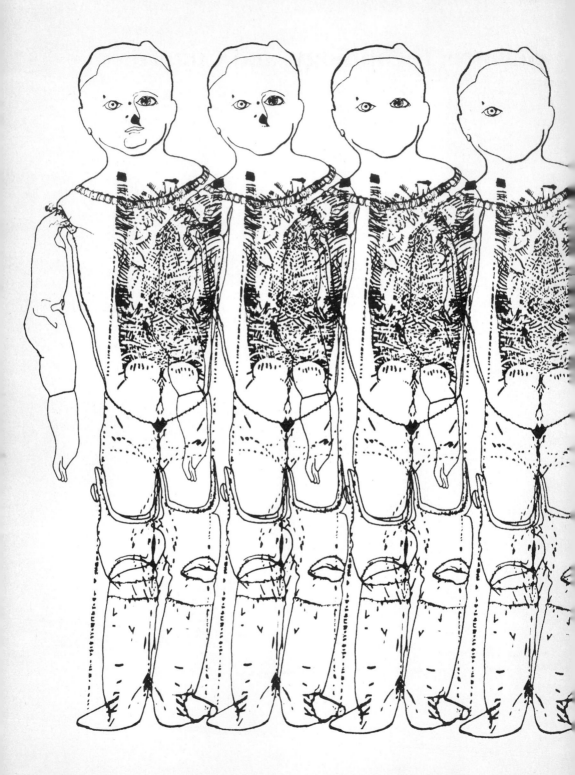

Science, Technology, and Culture

1

Introduction

The selections included in this chapter were chosen to illustrate the impact of modern science and technology on culture. By culture we mean the values, customs, institutions, and practices of a civilization; in this case, primarily recent and contemporary European and American civilization. This impact has been complex and because of the complexity and widespread changes introduced in all spheres of life, many people now view the future with uncertainty and fear. The cultural changes brought about by science and technology have been negative as well as positive. Let us briefly consider these effects in turn.

On the negative side there is great concern among our authors over the undermining of traditional humanist values. Life styles are imposed upon the majority of people. Fewer and fewer people have the opportunity to make their own decisions in any area. Work is no longer self-fulfilling even for the educated. Individual human needs are manipulated in the interest of profitable production; for example, instead of the buyer choosing the commodity, the commodity chooses the buyer. These alienating processes, made possible by advances in technology, are dehumanizing, and they are described and criticized in many of the following selections.

Scientific and technological advances are also blamed for continued breakdowns in social relations at work, at home, and at school. Specialization reinforces social fragmentation and cuts people off from one another. As each individual goes about the business of day-to-day living, we see the rise of a privileged technocratic leadership. These leaders, who possess special technical knowledge, are becoming a class of decision-makers who think and decide for everyone, even though their interests may be in conflict with those of the majority. The great masses of people are thus alienated from their leaders, and have no redress since they lack the required technical competence to run the society.

The processes of alienation and dehumanization are frequently attributed to the subordination of science and technology to vested interests, such as the military, big corporations, or a combination of the two, known as the military-industrial complex. These institutions are discussed in several articles of this chapter, and the authors agree that economic and military domination of science and technology is not in the best interests of the people.

It has been suggested that science and technology have destructive effects on the cultural fabric because they are internally rationalized, but externally and programmatically aimless, unplanned, and irrational. Technology is seen as pure technique; technique for the sake of technique, with no overriding purpose or rationale. What is feasible is put into effect, because what is feasible is mistakenly thought to be what is desirable. Thus economic and technical growth and expansion continue against the best interests of society even though they may be causing irreparable damage. Environmental pollution, which will be discussed in the last chapter of this book, is an outstanding example of this. Finally, the extension of scientific and technological procedures to the realm of human affairs is viewed by many authors in this chapter as guaranteeing the loss of human freedoms by supplying ever-newer, and ever-more effective techniques for controlling human behavior.

On the positive side, however, science and technology have been constructive and life-enhancing in many ways, and this is emphasized in the readings. Because of advanced scientific knowledge and application, the average standard of living has been remarkably increased and enhanced. Public health is greatly improved—many horrible and once common diseases have been eradicated. Literacy approaches 100 percent in all advanced countries. Developments in communication and transport are phenomenal. Human misery and suffering have been greatly alleviated. Outmoded values and practices have been largely abandoned. Superstition, racism, and bigotry have been lessened as a result of scientific knowledge. The problem seems to be that many of these positive accomplishments are outweighed by the negative forces; or they are suppressed and neglected as the scientific and technical apparatus is increasingly subjected to special interests.

Perhaps the principal question raised by the contributors to this discussion is, Can science and technology be liberated from subservience to special interests (economic, military, ideological, or a combination of the three) when these interests are inconsistent with the needs of society?

There is the hint of another possible interpretation in at least one of the selections. It raises the following questions: Are science and technology repressive and destructive in nature, by definition? Does science when applied on a large scale inevitably lend itself to exploitation, domination and destruction?

WILLIS H. TRUITT

Science, History, and Human Values

Willis H. Truitt is Associate Professor of Philosophy at the University of South Florida. He has contributed papers and edited books in the fields of history of science, aesthetics, social philosophy, and the philosophy of culture. Here Truitt considers the possibility of science as ideology, and the related questions: Is science free? Is it subservient to extra-scientific interests? In what sense are science and technology "repressive"? He also points out the paradoxes involved in any discussion about the social function of science.

Animals . . . change external nature by their activities just as man does, if not to the same extent, and these changes made by them in their environment . . . in turn react upon and change their originators. For in nature nothing takes place in isolation. Everything affects every other thing and *vice versa*, and it is usually because this many-sided motion and interaction is forgotten that our natural scientists are prevented from clearly seeing the simplest things.[1]

Previously unpublished.

Man's most powerful creation, his science, that which he himself fashioned, reacts back upon him and changes him, changes the nature of his life. But why do some believe that this change is rapidly becoming counter-adaptive, why is it frequently seen as a threat to man's future, why is it that science and technology are often condemned as sources of repression and destruction, rather than means to freedom, the good life and survival? The great optimism that accompanied the birth and early development of science has given way in our time to skepticism and cynicism about the role that science can play in bettering the human condition. Let us examine some of the reasons for this general change in men's attitudes toward science and technology.

There is no doubt that in the shorter view, man's scientific and technological achievements have served him well, that is, they have proved to be both productive, culturally enhancing and socially adaptive. And when in subsequent passages we refer to science as normative, we refer to those tendencies, functions, and factors which have had definite life-enhancing value. Thus there is a normative element, a prescriptive element, in all of human activity. And this normative or adaptive element is discoverable through a social and historical study of the activity in question. In the case of science we would want to ask, What human purposes has science been used to realize?

Again, paraphrasing Engels, no human activity takes place in isolation; there is no decontextualized act. All activity takes effect in the world and is reflected back upon the actor, the initiator. One might suppose that the determination of which activities are adaptive and which are either superfluous or nonadaptive requires only that we agree upon the activities, institutions and methods which have proved to be socially beneficial, productive, and highly advantageous in the past and present. Once these are agreed upon then we can employ them as guiding principles or outlines for a planned future social development. In this way we can learn to promote activities and institutions that are progressive and positive and reject tendencies that are clearly regressive and debasing. This then would seem to be the best way to evaluate the significance and value of science and technology.

At first sight scientific procedures and discoveries appear to be good. Science and technology are positive and progressive. Science and technology have provided the means for eradicating disease, poverty, and ignorance. But here we run into a paradox. For even though science has provided the means for abolishing a great deal of unnecessary human suffering, the suffering continues, and even increases during times of war or high unemployment, in the ghettos of our inner cities, in peripheral areas of the economy (the Appalachian coal mining region) and in certain occupations (migrant farm labor). Are science and technology incapable of solving these problems? Do science and technology actually perpetuate these problems (unemployment caused by automation, environmental pollution)? Is science neutral to such problems?

Professor Marx Wartofsky[2] has raised three questions which are highly relevant to this discussion:

1. Is value amenable to scientific study, and may the object of such study be taken as either natural, or human or societal fact? Is there a science of value?

2. What values are exhibited IN science? This is a question concerning the sociology of science or the study of the *ethos* of science [the ethic of science].

3. What is the value of science? What larger interests does it subserve or subvert?

As to the first question, the answer would seem to be implicit in our earlier remarks. What is value? Value is valuing behavior. It is not a transcendent entity, not supraempirical in any sense. It is inextricably related to human activities and cannot be studied independently of them. But accepted, established values and value systems are not necessarily humanly and socially beneficial merely because they have gained relatively wide acceptance—we need only look to history to make this point painfully clear.

The answer to Wartofsky's first question begins to come into focus. Value (valuing) like all forms of human, social behavior is a natural, human, social fact. It is therefore the subject of scientific study. The science of values will be concerned with the investigation of human behavior and institutions, and the further determination of the adaptive and meliorative achievements and potentials of such behavior and institutions. And this determination will, in turn, be based upon a careful and rigorous examination of social facts and cultural trends which will indicate in general those practices which, from the data of contemporary sociology and sociological history, appear most socially and adaptively beneficial. Science as a form of human activity and as an institution then must be subjected to such a study. What do we discover if we do this? The result seems paradoxical again and leads to the second question (What values are exhibited IN science, what is the *ethos* of science?). This question is more difficult to assess because of the paradoxical nature of the scientific enterprise itself. Robert Cohen has expressed two very antithetical approaches to the problem:

The full truth is bitter. Science is no longer the enlightening ally of human progress that it once seemed to be, and humane men will look warily at any model of a scientifically rationalized social order, at too strict a devotion to facts, at too concentrated a focus of intellectual resources upon the very technical fields which have enabled the mechanization of human life and culture. . . . We come to realize again that science is morally neutral. It has not automatically been a force for good. . . . Furthermore, the extension of science to the study of society and history is no guarantee of humane commitment within the scientific community, nor of moral wisdom within scientific knowledge.[3]

An opposite point of view, however, is contained in the following remarks:

In science, we combine subjective attitudes with objective demands; for example, an esthetic delight with a demanding reasonableness. We combine beauty with utility. We combine pride with modesty. We combine authority and leadership with private judgment and constant individual criticism. And we treat each other with respect. Despite violation by pride and other weaknesses, the ethic of the international community of scientists is known and persists. And the ethic of science is the democratic ethic of a cooperative republic.[4]

If we take the second of these opposed points of view as the true ethos of science—as the value system implicit in the scientific enterprise, then the answer to Wartofsky's third question (What is the value of science?) is made evident. It is that the function of science, its value, lies in its capacity to extend its own values to society and the world at large by means of social engineering. Science can transform the environment to meet human needs which have been scientifically and objectively established by the criteria and methods suggested above. But if we take the more skeptical analysis of science we seem to come up empty-handed— the paradox is reinforced. Which is the more accurate description of science?

Clearly not all people and not all scientists are willing to accept the humanistic or humane concept of science. Why do they not accept this vision of the scientific enterprise? Why should they accept it? To argue that the humanistic conception of science was implicit in the Greek conception of science is no justification for accepting it. Many of the authors in Chapter 2 make a great deal of this tendency in early Greek science and in the birth of modern science. But this interpretation of early scientific development does not provide an answer to why many scientists and laymen today are skeptical about the potential of science and technology to solve human social problems. Why are scientists reluctant to embrace a humanistic science?

The first argument against the acceptance of a valuative science, a normative science, is an analytic argument. It states briefly that the realms of value and fact are incommensurable, that science is entirely hypothetical and descriptive and cannot legitimately enter into the realm of morality. Those who attempt to empirically relate fact and value are, on this analysis, guilty of committing the "naturalistic fallacy." If values exist at all, they are said to be divine or transcendental ideals; nonnatural properties; purely linguistic entities: emotive, ejaculative, or rhetorical expressions; or some combination of these. We will not be concerned with refuting this argument directly but rather with how this attitude might have emerged. My guess is that many scientists today fear that if science were subordinated to any set of values from the outside, no matter how humane these values appeared to be, the result would be the suppression of free scientific inquiry. And, indeed, there is historical evidence to support such a fear: The suppression of science by the church, or, in the Soviet Union, the subordination of biology (genetics) to a distorted ideology.

The second argument against a valuative science is historical rather than analytic, and has been touched upon above. It simply cites the numerous instances in which scientific discoveries, technological advances, etc., have exerted a deleterious effect on human beings and human societies.

Since science and valuing behavior are both presumably social activities seeking further and better specific adaptation, we must now consider why much of contemporary science has lost the originally conscious, adaptive and normative character which is to be found among the ancients and among the founders of modern science. Where and why did modern science and technology reject its normative, humanitarian ethos? This is a historical and sociological question. And if we can show how and why this development has taken place, then we shall be able to better understand the anti-humanistic biases of many scientists and laymen. Two important factors appear to have influenced the dehumanization of science: (1) the diminution of the classical concept of Reason, and (2) the reinforcement of this decline in the industrial revolution and the rise of capitalism. Let us examine these in turn.

The Greek conception of Reason involved a rationalistic and humanistic presupposition in which the subjective and the objective, value and fact, were combined in a world theory. Philosophy and science, theory and practice, were indistinguishable. Scientific activity, to the extent that it takes place, is, therefore, guided by Reason, by human and social needs; it functions for human purposes and ends. This is "... the concept of Reason, according to which man's intellectual faculties are ... capable of determining his own life and of determining, defining, and changing the universe."[5]

This conception of science, which is preeminently humanistic and adaptive, was overthrown during the scientific and industrial revolutions in the sixteenth and seventeenth centuries. In the scientific sphere, Galileo established a rational, infinite universe "organized and defined by science itself." The concrete empirical universe of Aristotle was transformed into a reified system of numbers. Scientific experience was no longer the same as human experience, nor were scientific values *necessarily* human values—individual sense experiences themselves (like secondary qualities, color, odor, taste) became subjective, illusory, unreal and irrational.

The new science [did] not elucidate the conditions and limits of its evidence, validity and method; it [did] not elucidate its inherent historical denominator. It [remained] unaware of its own foundation; and it [was] therefore unable to recognize its servitude; unable to free itself from the ends set and given to science by the pre-given empirical reality ... [Hence] ... Reason loses its philosophical power and its scientific right to define and project ideas and modes of Being beyond and against those established by prevailing reality. I say: "beyond" the empirical reality, not in any *metaphysical* but in a historical sense, namely, in the sense of projecting essentially different, historical alternatives.[6]

This development greatly benefited nascent industrial development and early capitalism. By providing it with new technical methods, the new science enabled the productive and technological explosions which brought European civilization into the present era. The new science was reinforced by the new economics, as capitalism was reciprocally supplied with ever newer and better technical instruments of environmental exploitation.[7] Science and technology divested of their classical, humanistic meaning fell into the hands of vested interests, the entrepreneur; science became humanly neutral, utilitarian, an instrument of the expanding economy and the explosive new mode of production. Science was thus "liberated" from universal human concerns; its logic and purpose became internal and instrumentally "rationalized"; it could now serve any and all special interests, and it did.

As Marcuse suggests,

The ideational realm of Galilean science [idealized, mathematical science] no longer includes the moral, esthetic, political Forms, the *Ideas* of Plato. And separated from this realm, science develops now as an "absolute" in the literal sense no matter how relative within its own realm it may be, absolved from its own, prescientific and nonscientific conditions and foundations.[8]

It is necessary here to recognize the extraordinary importance of the relation of capitalist industrialization to the new technology, because it was precisely at this juncture in history that science, technology, business, and labor become unified. As will be seen in other essays in Chapter 2, labor power, as it became equated with science and technology, now took on some of the characteristics of the new science. Just as in the case of science, labor became humanly neutral, morally aimless; work became its own ethic, and its own end. Under these conditions labor could now be manipulated (as "manpower," "manhours," like "horsepower"), bought and sold, by and for the technological means of production. Instead of being a human project, issuing from individual and personal needs and desires, work became a function of alien depersonalized motives and production demands, the accumulation of profits, subject to coercion, unemployment and starvation being its alternative. Hence, work lost its personal, human, productive meaning and became alienated. And only the reciprocal convergence and development of the new science and the new economy could have created such a condition. As a consequence, the major portion of each individual's life was subordinated to reasons, purposes, needs, and ends that were not his. Each human existence was transformed into a contingency; individual existence transformed into alienated existence; and as this process progressed (ironically in the name of "progress") as society and life became increasingly mechanized and dehumanized in factories and other controlled working conditions, so also did man's relation to society, to other men, and to his own soul. He began to suffer deprivation in all spheres of life, in all his relations and responses. These tendencies have become characteristic of all industrial social orders, communist as well as noncommunist, totalitarian and constitutional. We have attempted here only to sketch some of the historical roots of this phenomenon, a phenomenon in which the sensuous, common, empirical reality, is transformed by the new science into an abstract, mathematical form, thus at one and the same time representing that reality as abstract relations and also disguising and concealing it. In this process the abstract method of science became identified with the real; the abstract scientific methodology (mathematics) *is* the reality. The scientific mathematization of nature and, by means of technology, the mechanization of social relations ruptured man's essential, sensuous and material relation to prescientific natural and social reality. And science as pure method shows itself to be purely neutral, a set of conventions which severs its intrinsic relation with nature, natural processes, man, and society. Science is, therefore, at the call of any project, any program or motive; it is suitably pure and suitably amoral. Thus the most horrendous and feared contingencies can be mathematized and abstracted into innocuous statistics.[9] In this, *human* reality is quantified, men = numbers (men are abstractions). In a sense, science becomes what it seeks to avoid becoming, an ideology in the highest degree, and it leaves a void where the human purpose of labor and social life once held sway. Science dispenses with the qualitative content of reality; it becomes preoccupied with technically exploitable knowledge.[10] Again we see a grotesque kind of irony. Scientists who fought so hard to make science a neutral and value-free institution, by their very success, have created a technique which is now dominated by political and economic interests external to science. The allocation of social resources to scientific research and technological development is partisan. Even so, the more or less isolated or specialist concerns of the modern scientists are not to be repudiated and denounced for this reason, nor even regretted, because the very dehumanization, desocialization (freeing the enterprise from religion and metaphysics), of science in the sixteenth century provided the impetus for the technological revolution, and the possibility of plenitude which exists in the present; again a queer kind of paradox.

The question now, however, is one of whether society can continue if its science, its very method of adaptation, remains aloof and separated from, and unresponsive to, society as a whole, and correspondingly, through unreflective application of technology, threatens to subjugate and annihilate the race. In order to be beneficial, in order to meet collective needs, science must be rehumanized, resituated into a proper functional relation within society. But is rehumanization of science and technology possible if the values inherent in it are inimical to human freedom and social well-being? It is far too easy to reject this possibility.

Many sociological and functional studies of science have begun. Most recently this has been a result of the preoccupation with environmental pollution. There is a movement from within science itself to begin a process and program of rehumanization, to resurrect the *ethos* of science. In this movement, J.D. Bernal has advocated ". . . 'one science,' comprising the science of nature and humanity, . . . in practice as well as in theory."[11]

The ideology of modern science, he admits, is derived from capitalism, but this is at best a socially benign ideology, at worst a destructive and dehumanizing force. The widely proclaimed neutrality of science is no more than a reification of procedures which when thoroughly "neutralized" are capable of employment for any cause, humane or inhumane. This situation in the present day, according to Bernal, is both unnecessary and intolerable. Man invented science; he must now exercise his right to its control, for all of humanity. We have before us the real historical alternative of bringing science into the service of humanity, or allowing it to function neutrally, and most often in the interests of a select few persons and nations. As Bernal has pointed out:

. . . the frustration of science is a very bitter thing. It shows itself as disease, enforced stupidity, misery, thankless toil, and premature death for the great majority, and an anxious, grasping and futile life for the remainder. Science can change all this, but only science working with those social forces which understand its functions and which march to the same ends.[12]

The science of the industrial revolution, although revolutionizing the means of production, thus making possible the liberation of humanity from material want, redefined society in terms of a utilitarian principle.[13] And this principle necessitated the suppression of all nonproductive, "nonrationalized" institutions and forms of behavior. The new science separated from its social foundations pressed unswervingly toward greater mechanization, efficiency and the suppression of spontaneity in human, social life. And as long as science and modern technology "pay off" in material benefits to those who control it for personal, corporate, or national interest, the smaller the possibility of criticizing this tendency and turning science to new ends becomes.

In the past, man has succeeded in adapting to a "hostile" natural environment. The question now, however, is whether he can adapt to an environment of his own making. And there is no sure answer to this question. The solution to the problem, however, is clearly not a philosophical solution since philosophers have only interpreted a world demanding change. The final decision to take action is a political decision, not a theoretical one. We certainly cannot go back. The human cost of abandoning science and technology would be far too great. So this is no solution; it is at best a daydream and at worst an immoral, Luddite fantasy. But we are faced with an incredibly difficult problem. Can science and technology be liberated from the institutions of special interest which distort it? Or are the values, techniques, in a word the nature of science itself, repressive and destructive. Despite the cogent arguments which have been given by some in favor of the latter proposition, we believe that science can be liberated and placed in the service of humanity. But this will be no easy task.

Bibliography

1. Frederick Engels, *The Dialectics of Nature* (New York: International Publishers, 1940), pp. 289–290.
2. Marx W. Wartofsky, *The Conceptual Foundations of Scientific Thought* (New York: Macmillan, 1968), p. 404.
3. Robert S. Cohen, "Science and Ethics," a paper presented at the International Conference on Science and Technology, Herceg Novi, Yugoslavia, 1964. As quoted by Wartofsky, *ibid.*, p. 411.
4. *Ibid.*, p. 414.
5. Herbert Marcuse, "On Science and Phenomenology," *Boston Studies in the Philosophy of Science* (New York: Humanities Press, vol. II, 1965), p. 280.
6. *Ibid.*, p. 283.
7. See the contributions of E. Zilsel and B. Hessen to Chapter 2.
8. *Ibid.*, p. 284.
9. The statistical "overkill" and acceptable risk projections (in millions of deaths and casualties) of nuclear war contingency planning.
10. Compare Jürgen Habermas, *Knowledge and Human Interest* (Boston: Beacon Press, 1971).
11. J.D. Bernal, *Marx and Science* (New York: International Publishers, 1952), p. 35.
12. J.D. Bernal, *The Social Function of Science* (London: Routledge and Kegan Paul, 1946), p. xv.
13. Compare Herbert Marcuse, *Eros and Civilization* (Boston: Beacon Press, 1955).

JACQUES ELLUL

The Technological Society

Jacques Ellul is Professor of History of Law and Social History at the University of Bordeaux, France. He has an international reputation as philosopher of the technological society. His book *The Technological Society* (1964), from which this selection is taken, has become a classic. Here he projects the implications of science and technology into the year 2000, with special reference to educational and genetic technique. He sees a danger in this development which arises from the worshiping of technique as technique, in a word, manipulation. In his opinion the present thrust of modern science and technology is conceptually and morally aimless, barren, and platitudinous.

. . . A Look at the Year 2000. In 1960 the weekly *l'Express* of Paris published a series of extracts from texts by American and Russian scientists concerning society in the year 2000. As long as such visions were purely a literary concern of science-fiction writers and sensational journalists, it was possible to smile at them.[1] Now we have like works from Nobel Prize winners, members of the Academy of Sciences of Moscow, and other scientific notables whose qualifications are beyond dispute. The visions of these gentlemen put science fiction in the shade. By the year 2000, voyages to the moon will be commonplace; so will inhabited artificial satellites. All food will be completely synthetic. The world's population will have increased fourfold but will have been stabilized. Sea water and ordinary rocks will yield all the necessary metals. Disease, as well as famine, will have been eliminated; and there will be universal hygienic inspection and control. The problems of energy production will have been completely resolved. Serious scientists, it must be repeated, are the source of these predictions, which hitherto were found only in philosophic utopias.

The most remarkable predictions concern the transformation of educational methods and the problem of human reproduction. Knowledge will be accumulated in "electronic banks" and transmitted directly to the human nervous system by means of coded electronic messages. There will no longer be any need of reading or learning mountains of useless information; everything will be received and registered according to the needs of the moment. There will be no need of attention or effort. What is needed will pass directly from the machine to the brain without going through consciousness.

In the domain of genetics, natural reproduction will be forbidden. A stable population will be necessary, and it will consist of the highest human types. Artificial insemination will be employed. This, according to Muller, will "permit the introduction into a carrier uterus of an ovum fertilized *in vitro*, ovum and sperm . . . having been taken from persons representing the masculine ideal and the feminine ideal, respectively. The reproductive cells in question will preferably be those of persons dead long enough that a true perspective of their lives and works, free of all personal prejudice, can be seen. Such cells will be taken from cell banks and will represent the most precious genetic heritage of humanity . . . The method will have to be applied universally. If the people of a single country were to apply it intelligently and intensively . . . they would quickly attain a practically invincible level of superiority . . ." Here is a future Huxley never dreamed of.

Perhaps, instead of marveling or being shocked, we ought to reflect a little. A question no one ever asks when confronted with the scientific wonders of the future concerns the interim period. Consider, for example, the problems of automation, which will become acute in a very short time. How, socially, politically, morally, and humanly, shall we contrive to get there? How are the prodigious economic problems, for example, of unemployment, to be solved? And, in Muller's more distant utopia, how shall we force humanity to refrain from begetting children naturally? How shall we force them to submit to constant and rigorous hygienic controls? How shall man be persuaded to accept a radical transformation of his traditional modes of nutrition? How and where shall we relocate a billion and a half persons who today make their livings from agriculture and who, in the promised ultrarapid conversion of the next forty years, will become completely useless as cultivators of the soil? How shall we distribute such numbers of people equably over the surface of the earth, particularly if the promised fourfold increase in population materializes? How will we handle the control and occupation of outer space in order to provide a stable *modus vivendi*? How shall national boundaries be made to disappear? (One of the last two would be a necessity.) There are many other "hows," but they are conveniently left unformulated. When we reflect on the serious although relatively minor problems that were provoked by the industrial exploitation of coal and electricity, when we reflect that after a hundred and fifty years these problems are still not satisfactorily resolved, we are entitled to ask whether there are any solutions to the infinitely more complex "hows" of the next forty years. In fact, there is one and only one means to their solution, a world-wide totalitarian dictatorship which will allow technique its full scope and at the same time resolve the concomitant difficulties. It is not difficult to understand why the scientists and worshippers of technology prefer not to dwell on this solution, but rather to leap nimbly across the dull and uninteresting intermediary period and land squarely in the golden age. We might indeed ask ourselves if we will succeed in getting through the transition period at all, or if the blood and the suffering required are not perhaps too high a price to pay for this golden age.

If we take a hard, unromantic look at the golden age itself, we are struck with the incredible naïveté of these scientists. They say, for example, that they will be able to shape and reshape at will human emotions, desires, and thoughts and arrive scientifically at certain efficient, pre-established collective decisions. They claim they will be in a position to develop certain collective desires, to constitute certain homogeneous social units out of aggregates of individuals, to forbid men to raise their children, and even to persuade them to renounce having any.

At the same time, they speak of assuring the triumph of freedom and of the necessity of avoiding dictatorship at any price.[2] They seem incapable of grasping the contradiction involved, or of understanding that what they are proposing, even after the intermediary period, is in fact the harshest of dictatorships. In comparison, Hitler's was a trifling affair. That it is to be a dictatorship of test tubes rather than of hobnailed boots will not make it any less a dictatorship.

When our savants characterize their golden age in any but scientific terms, they emit a quantity of down-at-the-heel platitudes that would gladden the heart of the pettiest politician. Let's take a few samples. "To render human nature nobler, more beautiful, and more harmonious." What on earth can this mean? What criteria, what content, do they propose? Not many, I fear, would be able to reply. "To assure the triumph of peace, liberty, and reason." Fine words with no substance behind them. "To eliminate cultural lag." What culture? And would the culture they have in mind be able to subsist in this harsh social organization? "To conquer outer space." For what purpose? The conquest of space seems to be an end in itself, which dispenses with any need for reflection.

We are forced to conclude that our scientists are incapable of any but the emptiest platitudes when they stray from their specialties. It makes one think back on the collection of mediocrities accumulated by Einstein when he spoke of God, the state, peace, and the meaning of life. It is clear that Einstein, extraordinary mathematical genius that he was, was no Pascal; he knew nothing of political or human reality, or, in fact, anything at all outside his mathematical reach. The banality of Einstein's remarks in matters outside his specialty is as astonishing as his genius within it. It seems as though the specialized application of all one's faculties in a particular area inhibits the consideration of things in general. Even J. Robert Oppenheimer, who seems receptive to a general culture, is not outside this judgment. His political and social declarations, for example, scarcely go beyond the level of those of the man in the street. And the opinions of the scientists quoted by l'Express are not even on the level of Einstein or Oppenheimer. Their pomposities, in fact, do not rise to the level of the average. They are vague generalities inherited from the nineteenth century, and the fact that they represent the furthest limits of thought of our scientific worthies must be symptomatic of arrested development or of a mental block. Particularly disquieting is the gap between the enormous power they wield and their critical ability, which must be estimated as null. To wield power well entails a certain faculty of criticism, discrimination, judgment, and option. It is impossible to have confidence in men who apparently lack these faculties. Yet it is apparently our fate to be facing a "golden age" in the power of sorcerers who are totally blind to the meaning of the human adventure. When they speak of preserving the seed of outstanding men, whom, pray, do they mean to be the judges. It is clear, alas, that they propose to sit in judgment themselves. It is hardly likely that they will deem a Rimbaud or a Nietzsche worthy of posterity. When they announce that they will conserve the genetic mutations which appear to them most favorable, and that they propose to modify the very germ cells in order to produce such and such traits; and when we consider the mediocrity of the scientists themselves outside the confines of their specialties, we can only shudder at the thought of what they will esteem most "favorable."

None of our wise men ever pose the question of the end of all their marvels. The "wherefore" is resolutely passed by. The response which would occur to our contemporaries is: for the sake of happiness. Unfortunately, there is no longer any question of that. One of our best-known specialists in diseases of the nervous system writes: "We will be able to modify man's emotions, desires and thoughts, as we have already done in a rudimentary way with tranquillizers." It will be possible, says our specialist, to produce a conviction or an impression of happiness without any real basis for it. Our man of the golden age, therefore, will be capable of "happiness" amid the worst privations. Why, then, promise us extraordinary comforts, hygiene, knowledge, and nourishment if, by simply manipulating our nervous systems, we can be happy without them? The last meager motive we could possibly ascribe to the technical adventure thus vanishes into thin air through the very existence of technique itself.

But what good is it to pose questions of motives? of Why? All that must be the work of some miserable intellectual who balks at technical progress. The attitude of the scientists, at any rate, is clear. Technique exists because it is technique. The golden age will be because it will be. Any other answer is superfluous.

Bibliography

1. Some excellent works, such as Robert Jungk's *Le Futur a déjà commencé*, were included in this classification.
2. The material here is cited from actual texts.

ERICH FROMM

The Revolution of Hope

Erich Fromm is an internationally known psychoanalyst. He is currently Professor of Psychoanalysis at the Medical School of the National University of Mexico. He has written widely in the fields of psychiatry, philosophy, social science, and religion. Among his many books are *Escape from Freedom* (1941), *Beyond the Chains of Illusion* (1962), and *The Sane Society* (1955). Fromm takes up the challenge of inevitable dehumanization in the year 2000. If present trends continue, Fromm foresees not the realization of the aspirations of men but the beginning of a period in which man ceases to be human and is transformed into an unthinking and unfeeling machine. However, he suggests that it is possible to reverse this trend. The key to redirection lies in breaking the production-consumption mechanism which demands administrative totalitarianism and acceleration in the methods of behavioral control. It is not too late, according to Fromm, to combat repressive manipulation and technocratic dictatorship.

What is the kind of society and the kind of man we might find in the year 2000, provided nuclear war has not destroyed the human race before then?

If people knew the likely course which American society will take, many if not most of them would be so horrified that they might take adequate measures to permit changing the course. If people are not aware of the direction in which they are going, they will awaken when it is too late and when their fate has been irrevocably sealed. Unfortunately, the vast majority are not aware of where they are going. They are not aware that the new society toward which they are moving is as radically different from Greek and Roman, medieval and traditional industrial societies as the agricultural society was from that of the food gatherers and hunters. Most people still think in the concepts of the society of the first Industrial Revolution. They see that we have more and better machines than man had fifty years ago and mark this down as progress. They believe that lack of direct political oppression is a manifestation of the achievement of personal freedom. Their vision of the year 2000 is that it will be the full realization of the aspirations of man since the end of the Middle Ages, and they do not see that the year 2000 may be not the fulfillment and happy culmination of a period in which man struggled for freedom and happiness, but the beginning of a period in which man ceases to be human and becomes transformed into an unthinking and unfeeling machine.

It is interesting to note that the dangers of the new dehumanized society were already clearly recognized by intuitive minds in the nineteenth century, and it adds to the impressiveness of their vision that they were people of opposite political camps.[1]

A conservative like Disraeli and a socialist like Marx were practically of the same opinion concerning the danger to man that would arise from the uncontrolled growth of production and consumption. They both saw how man would become weakened by enslavement to the machine and his own ever-increasing cupidity. Disraeli thought the solution could be found by containing the power of the new bourgeoisie; Marx believed that a highly industrialized society could be transformed into a humane one, in which man and not material goods were the goal of all social efforts.[2] One of the most brilliant progressive thinkers of the last century, John Stuart Mill, saw the problem with all clarity:

I confess I am not charmed with the ideal of life held out by those who think that the normal state of human beings is that of struggling to get on; that the trampling, crushing, elbowing, and treading on each other's heels, which form the existing type of social life, are the most desirable lot of human kind, or anything but the disagreeable symptoms of one of the phases of industrial progress. . . . Most fitting, indeed, is it, that while riches are power, and to grow as rich as possible the universal object of ambition, the path to its attainment should be open to all, without favor or partiality. But the best state for human nature is that in which, while no one is poor, no one desires to be richer, nor has any reason to fear being thrust back by the efforts of others to push themselves forward.[3]

It seems that great minds a hundred years ago saw what would happen today or tomorrow, while we to whom it is happening blind ourselves in order not to be disturbed in our daily routine. It seems that liberals and conservatives are equally blind in this respect. There are only few writers of vision who have clearly seen the monster to which we are giving birth. It is not Hobbes' *Leviathan*, but a Moloch, the all-destructive idol, to which human life is to be sacrificed. This Moloch has been described most imaginatively by Orwell and Aldous Huxley, by a number of science-fiction writers who show more perspicacity than most professional sociologists and psychologists.

I have already quoted Brzezinski's description of the technetronic society, and only want to quote the following addition: "The largely humanist-oriented, occasionally ideologically-minded intellectual-dissenter . . . is rapidly being displaced either by experts and specialists . . . or by the generalists-integrators, who become in effect house-ideologues for those in power, providing overall intellectual integration for disparate actions."[4]

A profound and brilliant picture of the new society has been given recently by one of the most outstanding humanists of our age, Lewis Mumford.[5] Future historians, if there are any, will consider his work to be one of the prophetic warnings of our time. Mumford gives new depth and perspective to the future by analyzing its roots in the past. The central phenomenon which connects past and future, as he sees it, he calls the "megamachine."

The "megamachine" is the totally organized and homogenized social system in which society as such functions like a machine and men like its parts. This kind of organization by total coordination, by "the constant increase of order, power, predictability and above all control," achieved almost miraculous technical results in early megamachines like the Egyptian and Mesopotamian societies, and it will find its fullest expression, with the help of modern technology, in the future of the technological society.

Mumford's concept of the megamachine helps to make clear certain recent phenomena. The first time the megamachine was used on a large scale in modern times was, it seems to me, in the Stalinist system of industrialization, and after that, in the system used by Chinese Communism. While Lenin and Trotsky still hoped that the Revolution would eventually lead to the mastery of society by the individual, as Marx had visualized, Stalin betrayed whatever was left of these hopes and sealed the betrayal by the physical extinction of all those in whom the hope might not have completely disappeared. Stalin could build his megamachine on the nucleus of a well-developed industrial sector, even though one far below those of countries like England or the United States. The Communist leaders in China were confronted with a different situation. They had no industrial nucleus to speak of. Their only capital was the physical energy and the passions and thoughts of 700 million people. They decided that by means of the complete coordination of this human material they could create the equivalent of the original accumulation of capital necessary to achieve a technical development which in a relatively short time would reach the level of that of the West. This total coordination had to be achieved by a mixture of force, personality cult, and indoctrination which is in contrast to the freedom and individualism Marx had foreseen as the essential elements of a socialist society. One must not forget, however, that the ideals of the overcoming of private egotism and of maximal consumption have remained elements in the Chinese system, at least thus far, although blended with totalitarianism, nationalism, and thought control, thus vitiating the humanist vision of Marx.

The insight into this radical break between the first phase of industrialization and the second Industrial Revolution, in which society itself becomes a vast machine, of which man is a living particle, is obscured by certain important differences between the megamachine of Egypt and that of the twentieth century. First of all, the labor of the live parts of the Egyptian machine was forced labor. The naked threat of death or starvation forced the Egyptian worker to carry out his task. Today, in the twentieth century, the worker in the most developed industrial countries, such as the United States, has a comfortable life—one which would have seemed like a life of undreamed-of luxury to his ancestor working a hundred years ago. He has, and in this point lies one of the errors of Marx, participated in the economic progress of capitalist society, profited from it, and, indeed, has a great deal more to lose than his chains.

The bureaucracy which directs the work is very different from the bureaucratic elite of the old megamachine. Its life is guided more or less by the same middle-class virtues that are valid for the worker; although its members are better paid than the worker, the difference in consumption is one of quantity rather than quality. Employers and workers smoke the same cigarettes and they ride in cars that look the same even though the better cars run more smoothly than the cheaper ones. They watch the same movies and the same television shows, and their wives use the same refrigerators.[6]

The managerial elite are also different from those of old in another respect: they are just as much appendages of the machine as those whom they command. They are just as alienated, or perhaps more so, just as anxious, or perhaps more so, as the worker in one of their factories. They are bored, like everyone else, and use the same antidotes against boredom. They are not as the elites were of old—a culture-creating group. Although they spend a good deal of their money to further science and art, as a class they are as much consumers of this "cultural welfare" as its recipients. The culture-creating group lives on the fringes. They are creative scientists and artists, but it seems that, thus far, the most beautiful blossom of twentieth-century society grows on the tree of science, and not on the tree of art.

The technetronic society may be the system of the future, but it is not yet here; it can develop from what is already here, and it probably will, unless a sufficient number of people see the danger and redirect our course. In order to do so, it is necessary to understand in greater detail the operation of the present technological system and the effect it has on man.

What are the guiding principles of this system as it is today?

It is programed by two principles that direct the efforts and thoughts of everyone working in it: The first principle is the maxim that something *ought* to be done because it is technically *possible* to do it. If it is possible to build nuclear weapons, they must be built even if they might destroy us all. If it is possible to travel to the moon or to the planets, it must be done, even if at the expense of many unfulfilled needs here on earth. This principle means the negation of all values which the humanist tradition has developed. This tradition said that something should be done because it is needed for man, for his growth, joy, and reason, because it is beautiful, good, or true. Once the principle is accepted that something ought to be done because it is technically possible to do it, all other values are dethroned, and technological development becomes the foundation of ethics.[7]

The second principle is that of *maximal efficiency and output*. The requirement of maximal efficiency leads as a consequence to the requirement of minimal individuality. The social machine works more efficiently, so it is believed, if individuals are cut down to purely quantifiable units whose personalities can be expressed on punched cards. These units can be administered more easily by bureaucratic rules because they do not make trouble or create friction. In order to reach this result, men must be de-individualized and taught to find their identity in the corporation rather than in themselves.

The question of economic efficiency requires careful thought. The issue of being economically efficient, that is to say, using the smallest possible amount of resources to obtain maximal effect, should be placed in a historical and evolutionary context. The question is obviously more important in a society where real material scarcity is the prime fact of life, and its importance diminishes as the productive powers of a society advance.

A second line of investigation should be a full consideration of the fact that efficiency is only a known element in already existing activities. Since we do not know much about the efficiency or inefficiency of untried approaches, one must be careful in pleading for things as they are on the grounds of efficiency. Furthermore, one must be very careful to think through and specify the area and time period being examined. What may appear efficient by a narrow definition can be highly inefficient if the time and scope of the discussion are broadened. In economics there is increasing awareness of what are called "neighborhood effects"; that is, effects that go beyond the immediate activity and are often neglected in considering benefits and costs. One example would be evaluating the efficiency of a particular industrial project only in terms of the immediate effects on this enterprise—forgetting, for instance, that waste materials deposited in nearby streams and the air represent a costly and a serious inefficiency with regard to the community. We need to clearly develop standards of efficiency that take account of time and society's interest as a whole. Eventually, the human element needs to be taken into account as a basic factor in the system whose efficiency we try to examine.

Dehumanization in the name of efficiency is an all-too-common occurrence; e.g., giant telephone systems employing Brave New World techniques of recording operators' contacts with customers and asking customers to evaluate workers' performance and attitudes, etc.—all aimed at instilling "proper" employee attitude, standardizing service, and increasing efficiency. From the narrow perspective of immediate company purposes, this may yield docile, manageable workers, and thus enhance company efficiency. In terms of the employees, as human beings, the effect is to engender feelings of inadequacy, anxiety, and frustration, which may lead to either indifference or hostility. In broader terms, even efficiency may not be served, since the company and society at large doubtless pay a heavy price for these practices.

Another general practice in organizing work is to constantly remove elements of creativity (involving an element of risk or uncertainty) and group work by dividing and subdividing tasks to the point where no judgment or interpersonal contact remains or is required. Workers and technicians are by no means insensitive to this process. Their frustration is often perceptive and articulate, and comments such as "We are human" and "The work is not fit for human beings" are not uncommon. Again, efficiency in a narrow sense can be demoralizing and costly in individual and social terms.

If we are only concerned with input-output figures, a system may give the impression of efficiency. If we take into account what the given methods do to the human beings in the system, we may discover that they are bored, anxious, depressed, tense, etc. The result would be a twofold one: (1) Their imagination would be hobbled by their psychic pathology, they would be uncreative, their thinking would be routinized and bureaucratic, and hence they would not come up with new ideas and solutions which would contribute to a more productive development of the system; altogether, their energy would be considerably lowered. (2) They would suffer from many physical ills, which are the result of stress and tension; this loss in health is also a loss for the system. Furthermore, if one examines what this tension and anxiety do to them in their relationship to their wives and children, and in their functioning as responsible citizens, it may turn out that for the system as a whole the seemingly efficient method is most inefficient, not only in human terms but also as measured by merely economic criteria.

To sum up: efficiency is desirable in any kind of purposeful activity. But it should be examined in terms of the larger systems, of which the system under study is only a part; it should take account of the human factor within the system. Eventually efficiency as such should not be a *dominant* norm in any kind of enterprise.

The other aspect of the same principle, that of *maximum output*, formulated very simply, maintains that the more we produce of whatever we produce, the better. The success of the economy of the country is measured by its rise of total production. So is the success of a company. Ford may lose several hundred million dollars by the failure of a costly new model, like the Edsel, but this is only a minor mishap as long as the production curve rises. The growth of the economy is visualized in terms of ever-increasing production, and there is no vision of a limit yet where production may be stabilized. The comparison between countries rests upon the same principle. The Soviet Union hopes to surpass the United States by accomplishing a more rapid rise in economic growth.

Not only industrial production is ruled by the principle of continuous and limitless acceleration. The educational system has the same criterion: the more college graduates, the better. The same in sports: every new record is looked upon as progress. Even the attitude toward the weather seems to be determined by the same principle. It is emphasized that this is "the hottest day in the decade," or the coldest, as the case may be, and I suppose some people are comforted for the inconvenience by the proud feeling that they are witnesses to the record temperature. One could go on endlessly giving examples of the concept that constant increase of quantity constitutes the goal of our life; in fact, that it is what is meant by "progress."

Few people raise the question of *quality*, or what all this increase in quantity is good for. This omission is evident in a society which is not centered around man any more, in which one aspect, that of quantity, has choked all others. It is easy to see that the predominance of this principle of "the more the better" leads to an imbalance in the whole system. If all efforts are bent on doing *more*, the quality of living loses all importance, and activities that once were means become ends.[8]

If the overriding economic principle is that we produce more and more, the consumer must be prepared to want—that is, to consume—more and more. Industry does not rely on the consumer's spontaneous desires for more and more commodities. By building in obsolescence it often forces him to buy new things when the old ones could last much longer. By changes in styling of products, dresses, durable goods, and even food, it forces him psychologically to buy more than he might need or want. But industry, in its need for increased production, does not rely on the consumer's needs and wants but to a considerable extent on advertising, which is the most important offensive against the consumer's right to know what he wants. The spending of 16.5 billion dollars on direct advertising in 1966 (in newspapers, magazines, radio, TV) may sound like an irrational and wasteful use of human talents, of paper and print. But it is not irrational in a system that believes that increasing production and hence consumption is a vital feature of our economic system, without which it would collapse. If we add to the cost of advertising the considerable cost for restyling of durable goods, especially cars, and of packaging, which partly is another form of whetting the consumer's appetite, it is clear that industry is willing to pay a high price for the guarantee of the upward production and sales curve.

The anxiety of industry about what might happen to our economy if our style of life changed is expressed in this brief quote by a leading investment banker:

Clothing would be purchased for its utility; food would be bought on the basis of economy and nutritional value; automobiles would be stripped to essentials and held by the same owners for the full 10 or 15 years of their useful lives; homes would be built and maintained for their characteristics of shelter, without regard to style or neighborhood. And what would happen to a market dependent upon new models, new styles, new ideas?[9]

What is the effect of this type of organization on man? It reduces man to an appendage of the machine, ruled by its very rhythm and demands. It transforms him into *Homo consumens*, the total consumer, whose only aim is to *have* more and to *use* more. This society produces many useless things, and to the same degree many useless people. Man, as a cog in the production machine, becomes a thing, and ceases to be human. He spends his time doing things in which he is not interested, with people in whom he is not interested, producing things in which he is not interested; and when he is not producing, he is consuming. He is the eternal suckling with the open mouth, "taking in," without effort and without inner activeness, whatever the boredom-preventing (and boredom-producing) industry forces on him—cigarettes, liquor, movies, television, sports, lectures—limited only by what he can afford. But the boredom-preventing industry, that is to say, the gadget-selling industry, the automobile industry, the movie industry, the television industry, and so on, can only succeed in preventing the boredom from becoming conscious. In fact, they increase the boredom, as a salty drink taken to quench the thirst increases it. However unconscious, boredom remains boredom nevertheless.

The passiveness of man in industrial society today is one of his most characteristic and pathological features. He takes in, he wants to be fed, but he does not move, initiate, he does not digest his food, as it were. He does not reacquire in a productive fashion what he inherited, but he amasses it or consumes it. He suffers from a severe systemic deficiency, not too dissimilar to that which one finds in more extreme forms in depressed people.

Man's passiveness is only one symptom among a total syndrome, which one may call the "syndrome of alienation." Being passive, he does not relate himself to the world actively and is forced to submit to his idols and their demands. Hence, he feels powerless, lonely, and anxious. He has little sense of integrity or self-identity. Conformity seems to be the only way to avoid intolerable anxiety—and even conformity does not always alleviate his anxiety.

No American writer has perceived this dynamism more clearly than Thorstein Veblen. He wrote:

In all the received formulations of economic theory, whether at the hands of the English economists or those of the continent, the human material with which the inquiry is concerned is conceived in hedonistic terms; that is to say, in terms of a passive and substantially inert and immutably given human nature. . . . The hedonistic conception of man is that of a lightning calculator of pleasures and pains, who oscillates like a homogeneous globule of desire of happiness under the impulse of stimuli that shift him about the area, but leave him intact. He has neither antecedent nor consequent. He is an isolated, definitive human datum, in stable equilibrium except for the buffets of the impinging forces that displace him in one direction or another. Self-imposed in elemental space, he spins symmetrically about his own spiritual axis until the parallelogram of forces bears down upon him, whereupon he follows the line of the resultant. When the force of the impact is spent, he comes to rest, a self contained globule of desire as before. Spiritually, the hedonistic man is not a prime mover. *He is not the seat of a process of living, except in the sense that he is subject to a series of permutations enforced upon him by circumstances external and alien to him.*[10]

Aside from the pathological traits that are rooted in passiveness, there are others which are important for the understanding of today's pathology of normalcy. I am referring to the growing split of cerebral-intellectual function from affective-emotional experience; the split between thought from feeling, mind from the heart, truth from passion.

Logical thought is not rational if it is merely logical[11] and not guided by the concern for life, and by the inquiry into the total process of living in all its concreteness and with all its contradictions. On the other hand, not only thinking but also emotions can be rational. *"Le coeur a ses raisons que la raison ne connaît point,"* as Pascal put it. (The heart has its reasons which reason knows nothing of.) Rationality in emotional life means that the emotions affirm and help the person's psychic structure to maintain a harmonious balance and at the same time to assist its growth. Thus, for instance, irrational love is love which enhances the person's dependency, hence anxiety and hostility. Rational love is a love which relates a person intimately to another, at the same time preserving his independence and integrity.

Reason flows from the blending of rational thought and feeling. If the two functions are torn apart, thinking deteriorates into schizoid intellectual activity, and feeling deteriorates into neurotic life-damaging passions.

The split between thought and affect leads to a sickness, to a low-grade chronic schizophrenia, from which the new man of the technetronic age begins to suffer. In the social sciences it has become fashionable to think about human problems with no reference to the feelings related to these problems. It is assumed that scientific objectivity demands that thoughts and theories concerning man be emptied of all emotional concern with man.

An example of this emotion-free thinking is Herman Kahn's book on thermonuclear warfare. The question is discussed: how many millions of dead Americans are "acceptable" if we use as a criterion the ability to rebuild the economic machine after nuclear war in a reasonably short time so that it is as good as or better than before. Figures for GNP and population increase or decrease are the basic categories in this kind of thinking, while the question of the human results of nuclear war in terms of suffering, pain, brutalization, etc., is left aside.

Kahn's *The Year 2000* is another example of the writing which we may expect in the completely alienated megamachine society. Kahn's concern is that of the figures for production, population increase, and various scenarios for war or peace, as the case may be. He impresses many readers because they mistake the thousands of little data which he combines in ever-changing kaleidoscopic pictures for erudition or profundity. They do not notice the basic superficiality in his reasoning and the lack of the human dimension in his description of the future.

When I speak here of low-grade chronic schizophrenia, a brief explanation seems to be needed. Schizophrenia, like any other psychotic state, must be defined not only in psychiatric terms but also in social terms. Schizophrenic experience *beyond* a certain threshold would be considered a sickness in any society, since those suffering from it would be unable to function under any social circumstances (unless the schizophrenic is elevated into the status of a god, shaman, saint, priest, etc.). But there are low-grade chronic forms of psychoses which can be shared by millions of people and which—precisely because they do not go beyond a certain threshold—do not prevent these people from functioning socially. As long as they share their sickness with millions of others, they have the satisfactory feeling of not being alone; in other words, they avoid that sense of complete isolation which is so characteristic of full-fledged psychosis. On the contrary, they look at themselves as normal and at those who have not lost the link between heart and mind as being "crazy." In all low-grade forms of psychoses, the definition of sickness depends on the question as to whether the pathology is shared or not. Just as there is low-grade chronic schizophrenia, so there exist also low-grade chronic paranoia and depression. And there is plenty of evidence that among certain strata of the population, particularly on occasions where a war threatens, the paranoid elements increase but are not felt as pathological as long as they are common.[12]

The tendency to install technical progress as the highest value is linked up not only with our overemphasis on intellect but, most importantly, with a deep emotional attraction to the mechanical, to all that is not alive, to all that is man-made. This attraction to the non-alive, which is in its more extreme form an attraction to death and decay (necrophilia), leads even in its less drastic form to indifference toward life instead of "reverence for life." Those who are attracted to the non-alive are the people who prefer "law and order" to living structure, bureaucratic to spontaneous methods, gadgets to living beings, repetition to originality, neatness to exuberance, hoarding to spending. They want to control life because they are afraid of its uncontrollable spontaneity; they would rather kill it than to expose themselves to it and merge with the world around them. They often gamble with death because they are not rooted in life; their courage is the courage to die and the symbol of their ultimate courage is the Russian roulette.[13] The rate of our automobile accidents and the preparation for thermonuclear war are a testimony to this readiness to gamble with death. And who would not eventually prefer this exciting gamble to the boring unaliveness of the organization man?

One symptom of the attraction of the merely mechanical is the growing popularity, among some scientists and the public, of the idea that it will be possible to construct computers which are no different from man in thinking, feeling, or any other aspect of functioning.[14] The main problem, it seems to me, is not whether such a computer-man can be constructed; it is rather why the idea is becoming so popular in a historical period when nothing seems to be more important than to transform the existing man into a more rational, harmonious, and peace-loving being. One cannot help being suspicious that often the attraction of the computer-man idea is the expression of a flight from life and from humane experience into the mechanical and purely cerebral.

The possibility that we can build robots who are like men belongs, if anywhere, to the future. But the present already shows us men who act like robots. When the majority of men are like robots, then indeed there will be no problem in building robots who are like men. The idea of the manlike computer is a good example of the alternative between the human and the inhuman use of machines.

The computer can serve the enhancement of life in many respects. But the idea that it replaces man and life is the manifestation of the pathology of today.

The fascination with the merely mechanical is supplemented by an increasing popularity of conceptions that stress the animal nature of man and the instinctive roots of his emotions or actions. Freud's was such an instinctive psychology; but the importance of his concept of libido is secondary in comparison with his fundamental discovery of the unconscious process in waking life or in sleep. The most popular recent authors who stress instinctual animal heredity, like Konrad Lorenz (*On Aggression*) or Desmond Morris (*The Naked Ape*), have not offered any new or valuable insights into the specific human problem as Freud has done; they satisfy the wish of many to look at themselves as determined by instincts and thus to camouflage their true and bothersome human problems.[15] The dream of many people seems to be to combine the emotions of a primate with a computerlike brain. If this dream could be fulfilled, the problem of human freedom and of responsibility would seem to disappear. Man's feelings would be determined by his instincts, his reason by the computer; man would not have to give an answer to the questions his existence asks him. Whether one likes the dream or not, its realization is impossible, the naked ape with the computer brain would cease to be human, or rather "he" would not *be*.[16]

Among the technological society's pathogenic effects upon man, two more must be mentioned: the disappearance of *privacy* and of *personal human contact*.

"Privacy" is a complex concept. It was and is a privilege of the middle and upper classes, since its very basis, private space, is costly. This privilege, however, can become a common good with other economic privileges. Aside from this economic factor, it was also based on a hoarding tendency in which *my* private life was *mine* and nobody else's, as was *my* house and any other property. It was also a concomitant of *cant*, of the discrepancy between moral appearances and reality. Yet when all these qualifications are made, privacy still seems to be an important condition for a person's productive development. First of all, because privacy is necessary to collect oneself and to free oneself from the constant "noise" of people's chatter and intrusion, which interferes with one's own mental processes. If all private data are transformed into public data, experiences will tend to become more shallow and more alike. People will be afraid to feel the "wrong thing"; they will become more accessible to psychological manipulation which, through psychological testing, tries to establish norms for "desirable," "normal," "healthy" attitudes. Considering that these tests are applied in order to help the companies and government agencies to find the people with the "best" attitudes, the use of psychological tests, which is by now an almost general condition for getting a good job, constitutes a severe infringement on the citizen's freedom. Unfortunately, a large number of psychologists devote whatever knowledge of man they have to his manipulation in the interests of what the big organization considers efficiency. Thus, psychologists become an important part of the industrial and governmental system while claiming that their activities serve the optimal development of man. This claim is based on the rationalization that what is best for the corporation is best for man. It is important that the managers understand that much of what they get from psychological testing is based on the very limited picture of man which, in fact, management requirements have transmitted to the psychologists, who in turn give it back to management, allegedly as a result of an independent study of man. It hardly needs to be said that the intrusion of privacy may lead to a control of the individual which is more total and could be more devastating than what totalitarian states have demonstrated thus far. Orwell's 1984 will need much assistance from testing, conditioning, and smoothing-out psychologists in order to come true. It is of vital importance to distinguish between a psychology that understands and aims at the well-being of man and a psychology that studies man as an object, with the aim of making him more useful for the technological society. . . .

Bibliography

1. Cf. the statements by Burckhardt, Proudhon, Baudelaire, Thoreau, Marx, Tolstoy quoted in *The Sane Society* (New York: Rinehart, 1955), pp. 184 ff.
2. Cf. Erich Fromm, *Marx's Concept of Man* (New York: Ungar, 1961).
3. *Principles of Political Economy* (London: Longmans, 1929; 1st edition, 1848).
4. "The Technetronic Society," p. 19.
5. Lewis Mumford, *The Myth of the Machine* (New York: Harcourt, Brace & World, 1967).
6. The fact that the underdeveloped sector of the population does not take part in this new style of life has been mentioned above
7. While revising this manuscript I read a paper by Hasan Ozbekhan, "The Triumph of Technology: 'Can' Implies 'Ought.' " This paper, adapted from an invited presentation at MIT and published in mimeographed form by System Development Corporation, Santa Monica, California, was sent to me by the courtesy of Mr. George Weinwurm. As the title indicates, Ozbekhan expresses the same concept as the one I present in the text. His is a brilliant presentation of the problem from the standpoint of an outstanding specialist in the field of management science, and I find it a very encouraging fact that the same idea appears in the work of authors in fields as different as his and mine. I quote a sentence that shows the identity of his concept and the one presented in the text: "Thus, feasibility, which is a strategic concept, becomes elevated into a normative concept, with the result that whatever technological reality indicates we *can* do is taken as implying that we *must* do it" (p. 7).
8. I find in C. West Churchman's *Challenge to Reason* (New York: McGraw-Hill, 1968) an excellent formulation of the problem:
   "If we explore this idea of a larger and larger model of systems, we may be able to see in what sense completeness represents a challenge to reason. One model that seems to be a good candidate for completeness is called an *allocation* model; it views the world as a system of activities that use resources to "output" usable products.
   "The process of reasoning in this model is very simple. One searches for a central quantitative measure of system performance, which has the characteristic: the more of this quantity the better. For example, the more profit a firm makes, the better. The more qualified students a university graduates, the better. The more food we produce, the better. It will turn out that the particular choice of the measure of system performance is not critical, so long as it is a measure of general concern.
   "We take this desirable measure of performance and relate it to the feasible activities of the system. The activities may be the operations of various manufacturing plants, of

schools and universities, of farms, and so on. Each significant activity contributes to the desirable quantity in some recognizable way. The contribution, in fact, can often be expressed in a mathematical function that maps the amount of activity onto the amount of the desirable quantity. The more sales of a certain product, the higher the profit of a firm. The more courses we teach, the more graduates we have. The more fertilizer we use, the more food [pp. 156–57]."

9. Paul Mazur, *The Standards We Raise* (New York, 1953), p. 32.

10. "Why Is Economics Not an Evolutionary Science?," in *The Place of Science in Modern Civilization and Other Essays* (New York: B.W. Huebsch, 1919), p. 73. (Emphasis added.)

11. Paranoid thinking is characterized by the fact that it can be completely logical, yet lack any guidance by concern or concrete inquiry into reality; in other words, logic does not exclude madness.

12. The difference between that which is considered to be sickness and that which is considered to be normal becomes apparent in the following example. If a man declared that in order to free our cities from air pollution, factories, automobiles, airplanes, etc., would have to be destroyed, nobody would doubt that he was insane. But if there is a consensus that in order to protect our life, our freedom, our culture, or that of other nations which we feel obliged to protect, thermonuclear war might be required as a last resort, such opinion appears to be perfectly sane. The difference is not at all in the kind of thinking employed but merely in that the first idea is not shared and hence appears abnormal while the second is shared by millions of people and by powerful governments and hence appears to be normal.

13. Michael Maccoby has demonstrated the incidence of the life-loving versus the death-loving syndrome in various populations by the application of an "interpretative" questionnaire. Cf. his "Polling Emotional Attitudes in Relation to Political Choices" (to be published).

14. Dean E. Wooldridge, for instance, in *Mechanical Man* (New York: McGraw-Hill, 1968), writes that it will be possible to manufacture computers synthetically which are "completely undistinguishable from human beings produced in the usual manner"[!] (p. 172). Marvin L. Minsky, a great authority on computers, writes in his book *Computation* (Englewood Cliffs, N.J.: Prentice-Hall, 1967): "There is no reason to suppose machines have any limitations not shared by man" (p. vii).

15. This criticism of Lorenz refers only to that part of his work in which he deals by analogy with the psychological problems of man, not with his work in the field of animal behavior and instinct theory.

16. In revising this manuscript I became aware that Lewis Mumford had expressed the same idea in 1954 in *In the Name of Sanity* (New York: Harcourt, Brace & Co.): "Modern man, therefore, now approaches the last act of his tragedy, and I could not, even if I would, conceal its finality or its horror. We have lived to witness the joining, in intimate partnership, of the automaton and the id, the id rising from the lower depths of the unconscious, and the automaton, the machine-like thinker and the man-like machine, wholly detached from other life-maintaining functions and human reactions, descending from the heights of conscious thought. The first force has proved more brutal, when released from the whole personality, than the most savage of beasts; the other force, so impervious to human emotions, human anxieties, human purposes, so committed to answering only the limited range of questions for which its apparatus was originally loaded, that it lacks the saving intelligence to turn off its own compulsive mechanism, even though it is pushing science as well as civilization to its own doom [p. 198]."

J. D. BERNAL

Marx and Science

John Desmond Bernal was Professor of Physics and Crystallography at Birkbeck College, London University, from 1937 to 1968. He made important contributions to the theory of the origin of life. His major works include *The Social Function of Science* (1939), *The Freedom of Necessity* (1949), *Science in History* (1954), and *Origin of Life* (1967). The following selection by Bernal succinctly renders Marx's views on the relationship of science to economics and culture. We see that Marx very early recognized the machine as the prototype of modern technology, and that implicit in the machine are the abuses as well as the advantages it offers to man and society.

. . . Marx learned his economics at the center of the economic life of the world of his time, in England, and particularly in London and Manchester. As he acquired it, it gave him a greater grasp of all other aspects of culture. Marx's understanding of science and its relation to economic and social change was to continue to deepen all through his life and was enriched by the new experience of practical science and technology which he acquired in England. Of the pair, it was Engels who was more closely attached to the techniques of the productive process and to the general field of natural science:

Marx and I were pretty well the only people to rescue conscious dialectics from German idealist philosophy and apply it in the materialist conception of nature and history. But a knowledge of mathematics and natural science is essential to a conception of nature which is dialectical and at the same time materialist. Marx was well versed in mathematics, but we could only partially, intermittently and sporadically keep up with the natural sciences. For this reason, when I retired from business and transferred my home to London, thus enabling myself to give the necessary time to it, I went through as complete as possible a 'molting,' as Liebig calls it, in mathematics and the natural sciences, and spent the best part of eight years on it.

Nevertheless Marx himself worked hard at acquiring the necessary basic and even practical knowledge. For example he wrote to Engels:

I am adding something to the section on machinery.[1] There are some curious questions here which I ignored in my first treatment. In order to get clear about it I have read through all my notebooks (extracts) on technology again and am also attending a practical course (experimental only) for workers, by Professor Willis (at the Geological Institute in Jermyn Street, where Huxley also used to give his lectures). It is the same for me with mechanics as it is with languages. I understand the mathematical laws, but the simplest technical reality demanding perception is harder to me than to the biggest blockheads.

Marx never shone as a hard worker. At the depths of his financial difficulties he did get a job as a railway clerk, but was only able to keep it for a few days on account of his bad handwriting.

Through even closer association with Engels, Marx was able to see and analyze the actual processes of industry and to relate them in detail to their economic consequences. This is shown very clearly in his great work *Capital*, particularly in Chapter XV of the first volume, on "Machinery and Modern Industry," and in Chapter V of the third volume, on "Economies in the Employment of Constant Capital." The opening passages of the former are astonishing to read even today in their clarity and penetration. Marx showed an understanding of the essence of mechanical production which was far ahead of that of anyone else of his time. One only has to read the ideas of a very intelligent and penetrating English scientist, Charles Babbage,[2] to see the enormous advantage which Marx drew from his more comprehensive, philosophic and economic approach. Where Babbage only saw individual examples of the use of machinery, Marx could see a single continuous transforming process. This process started with the handicraftsman with his tools, moved on to the period which he called that of manufacture, where a number of handicrafts are put together and where a division of labor results in lowered costs, to reach the position of modern industry where the machine enters the field.

Marx first analyzes the machinery of productive industry in a general way:

All fully developed machinery consists of three essentially different parts, the motor mechanism, the transmitting mechanism, and finally the tool or working machine. The motor mechanism is that which puts the whole in motion. It either generates its own motive power, like the steam engine, the caloric engine, the electro-magnetic machine, etc., or it receives its impulse from some already existing natural force, like the water-wheel from a head of water, the wind-mill from wind, etc. The transmitting mechanism, composed of fly-wheels, shafting, toothed wheels, pullies, straps, ropes, bands, pinions, and gearing of the most varied kinds, regulates the motion, changes its form where necessary, as for instance, from linear to circular, and divides and distributes it among the working machines. These two first parts of the whole mechanism are there, solely for putting the working machines in motion, but means of which motion the subject of labor is seized upon and modified as desired. The tool or working-machine is that part of the machinery with which the industrial revolution of the 18th century started. And to this day it constantly serves as such a starting point, whenever a handicraft, or a manufacture, is turned into an industry carried on by machinery.

This leads him to consider the essential character of a machine to be the fact that it is a tool operated not by a man but by a mechanical contrivance:

The machine proper is therefore a mechanism that, after being set in motion, performs with its tools the same operations that were formerly done by the workman with similar tools. . . .

The machine, which is the starting point of the industrial revolution, supersedes the workman, who handles a single tool, by a mechanism operating with a number of similar tools, and set in motion by a single motive power, whatever the form of that power may be. Here we have the machine, but only as an elementary factor of production by machinery.

Increase in the size of the machine, and in the number of its working tools, calls for a more massive mechanism to drive it; and this mechanism requires, in order to overcome its resistance, a mightier moving power than that of man, apart from the fact that man is a very imperfect instrument for producing uniform continued motion. But assuming that he is acting simply as a motor, that a machine has taken the place of his tool, it is evident that he can be replaced by natural forces.

He saw the first phase of machine industry arising, not out of any radically new invention, but by the multiplication of simple handicraft operations, linked by such a mechanism as that of the spinning jenny or Crompton's mule. His analysis of the later stages of the development of industry was even more penetrating. He showed how it was changing: first, by the blending of different machines into each other to form more complex machines and leading the way towards the continuous flow, semi- or completely automatic process that we consider characteristic of twentieth-century industry; and secondly by the enlargement of mechanical means, so as to do things which were impossible by limited individual human strength, particularly in the heavy engineering and iron and steel industries:

Modern Industry had therefore itself to take in hand the machine, its characteristic instrument of production, and to construct machines by machines. It was not till it did this, that it built up for itself a fitting technical foundation, and stood on its own feet. Machinery, simultaneously with the increasing use of it, in the first decades of this century, appropriated, by degrees, the fabrication of machines proper. But it was only during the decade preceding 1866, that the construction of railways and ocean steamers on a stupendous scale called into existence the cyclopean machines now employed in the construction of prime movers.

He saw further that this development was linking science with industry and that it was to have far-reaching social consequences.

The implements of labor, in the form of machinery, necessitate the substitution of natural forces for human force, and the conscious application of science, instead of rule of thumb. In Manufacture, the organization of the social labor-process is purely subjective; it is a combination of detail laborers; in its machinery system, Modern Industry has a productive organism that is purely objective, in which the laborer becomes a mere appendage to an already existing material condition of production. In simple cooperation, and even in that founded on division of labor, the suppression of the isolated, by the collective, workman still appears to be more or less accidental. Machinery, with a few exceptions to be mentioned later, operates only by means of associated labor, or labor in common. Hence the cooperative character of the labor-process is, in the latter case, a technical necessity dictated by the instrument of labor itself.

Marx was able to have this functional understanding of machinery because he linked it at every stage with its actual economic use. He demonstrated that the reason for John Stuart Mill's complaint that machinery had not "lightened the day's toil of any human being" was that this had never been the motive of invention under capitalism. That motive had been first and last that of profit. The function of technical improvement was primarily to increase the value of the product for the same labor force, and secondarily to increase the rate of profit by increasing the quantity of raw materials worked up in a given period of the employment of plant and machinery. (See the discussion in *Capital*, Vol. III, Chapter V.) He further showed that, paradoxically, the more labor-saving the machinery the more people could profitably be brought in to work on it. The development of industry towards mass-production is very clearly foreshadowed in this part of his work.

Marx also understood well what science had to do in the development of modern industry. The demand for ever greater speed and economy of operation was one that rule of thumb improvement could no longer satisfy.

. . . cooperative and social production, a cooperation within the primary process of production. On the one hand, this is the indispensable requirement for the application of mechanical and chemical inventions without increasing the price of commodities, and this is always the first consideration. On the other hand, only production on a large scale permits those economies which are derived from cooperative productive consumption. Finally, it is only the experience of combined laborers which discovers the where and how of economies, the simplest methods of applying the experience gained, the way to overcome practical frictions in carrying out theories, etc.

Incidentally it should be noted that there is a difference between universal labor and cooperative labor. Both kinds play their role in the process of production, both flow one into the other, but both are also differentiated. *Universal labor is scientific labor, such as discoveries and inventions.* This labor is conditioned on the cooperation of living fellow-beings and on the labors of those who have gone before; cooperative labor, on the other hand, is a direct cooperation of living individuals.

The foregoing is corroborated by frequent observation, to-wit:

(1) The great difference in the cost of the first building of a new machine and that of its reproduction, on which see Ure and Babbage.

(2) The far greater cost of operating an establishment based on a new invention as compared to later establishments arising out of the ruins of the first one, as it were. This is carried to such an extent that the first leaders in a new enterprise are generally bankrupted, and only those who later buy the buildings, machinery, etc., cheaper, make money out of it. It is, therefore, generally the most worthless and miserable sort of money-capitalists who draw the greatest benefits out of the universal labor of the human mind and its cooperative application in society. [My italics—J.D.B.]

Thus he saw this universal labor, science, as a component of the productive force distinct from the older cooperative labor and, to a certain extent under capitalism, opposed to it. This is clearly stated in *Capital:*

It is a result of the division of labor in manufactures, that the laborer is brought face to face with the intellectual potencies of the material process of production, as the property of another, and as a ruling power. This separation

begins in simple cooperation, where the capitalist represents to the single
workman, the oneness and the will of the associated labor. It is developed in
manufacture which cuts down the laborer into a detail laborer. It is completed
in modern industry, which makes science a productive force distinct from labor
and presses it into the service of capital.

But if capitalism had built up science as a productive force, the very character
of the new mode of production was serving to make capitalism itself unnecessary.
Even while Marx was writing, in the very heyday of capitalism, he was able
to see signs of its decay and the beginning of the process of monopolistic
restriction that has grown so monstrously since his time. But Marx knew well
enough that however superfluous and even disastrous capitalism was becoming,
it would not vanish of itself. Nor would it merge imperceptibly into a better
system, as well-meaning or cowardly liberals or socialists would have liked to
think. He knew that the full social use of science could come only when the
proletariat, the class that had been called into existence by industry, itself
controlled the productive system that it was already maintaining by its own
co-operative labor. Marx said this plainly in the speech which he gave at the
anniversary dinner of the *People's Paper* in 1856:

There is one great fact, characteristic of this, our nineteenth century, a fact
which no party dares deny. On the one hand, there have started into life
industrial and scientific forces, which no epoch of the former human history
had ever suspected. On the other hand, there exist symptoms of decay, far
surpassing the horrors recorded of the latter times of the Roman Empire. In
our days everything seems pregnant with its contrary; machinery gifted with
the wonderful power of shortening and fructifying human labor, we behold
starving and overworking it. The new-fangled sources of wealth, by some strange
weird spell, are turned into sources of want. The victories of art seem bought
by the loss of character. At the same pace that mankind masters nature, man
seems to become enslaved to other men or to his own infamy. Even the pure
light of science seems unable to shine but on the dark background of ignorance.
All our invention and progress seem to result in endowing material forces
with intellectual life, and in stultifying human life into a material force. This
antagonism between modern industry and science on the one hand, modern
misery and dissolution on the other hand; this antagonism between the productive powers and the social relations of our epoch is a fact, palpable, overwhelming, and not to be controverted. Some parties may wail over it; others
may wish to get rid of modern arts in order to get rid of modern conflicts. Or
they may imagine that so signal a progress in industry wants to be completed
by as signal a regress in politics. On our part, we do not mistake the shape of
the shrewd spirit that continues to mark all these contradictions. We know that
to work well the new-fangled forces of society, they only want to be mastered
by new-fangled men—and such are the working men. They are as much the
invention of modern time as machinery itself. In the signs that bewilder the
middle class, the aristocracy and the poor prophets of regression, we do
recognize our brave friend, Robin Goodfellow, the old mole, that can work in
the earth so fast, that worthy pioneer—the revolution. The English working
men are the first born sons of modern industry. They will then, certainly, not

be the last in aiding the social revolution produced by that industry, a revolution, which means the emancipation of their own class all over the world, which is as universal as capital-rule and wages-slavery.

In this Marx brings out both the importance of science and the fact that it is only through the working class that it can effectively be used. The essential feature of modern industry as he saw it—the social production of value—cannot work effectively unless it is accompanied by the social utilization of the values produced. The only people who can ensure that social utilization are the people who suffer from the present system and who themselves are the major motive force of that system—the industrial workers.

Marx here clearly foreshadows a productive system which would be far more consciously controlled by the people than anything that capitalism could evolve. In this control he sees the possibility of achieving results which are impossible in the constant pursuit of profit, which warps all constructive enterprise, and in the anarchy of production imposed by the conditions of the market. That social control is, therefore, itself a condition of freedom. It was to this end that Marx called on the working class to take the matter into their own hands by overthrowing the bourgeois state. Then only would the "one science," comprising the science of nature and of humanity, be able to take form in practice as well as in theory. In this, as in everything else, in philosophy as well as politics, Marx throughout his whole life wove the future into the present. He both foresaw and ensured the realization of his prophecy.

Looking back now over the years since Marx's death we should be able to appreciate something of the importance of his understanding of the relations between science, production, and political forms. Yet how few intellectuals with their knowledge and the experience of the great and terrible events of our time have even begun to do so! Certainly the majority of intellectuals of his own time did not. Most of the "well educated" scientists who were, whether they liked it or not, components of the productive mechanism, the economists and philosophers paid to provide the ideological background of the capitalist system, were incapable of refuting Marxism because they were incapable of looking at it at all, much less of understanding it. The Marxist ideas spread among the working class, which was the only class capable of appreciating from the experience of their own lives the essential features of this philosophy, and particularly the need to combine at every stage their understanding with their action.

Marx himself had been the first to set out the laws of transformation of human society. From the very moment he did so he became an active working-class leader. It was this aspect of his activity, transmitted through an increasing class-conscious proletariat, that was to prove the effective means of carrying out the task of "changing the world" which he himself had given to the philosophers. Lafargue indeed had written:

Karl Marx was one of those rare men who are fitted for the front rank both in science and in public life. So intimately did he combine these two fields that we shall never understand him unless we regard him simultaneously as man of science and as socialist fighter. While he was of the opinion that every science must be cultivated for its own sake and that when we undertake scientific research we should not trouble ourselves about the possible consequences, nevertheless, he held that the man of learning, if he does not wish to degrade himself, must never cease to participate in public affairs—must not be content to shut himself up in his study or his laboratory, like a maggot in a cheese, and to shun the social and political struggles of his contemporaries. 'Science must not be a selfish pleasure. Those who are so lucky as to be able to devote themselves to scientific pursuits should be the first to put their knowledge at the service of mankind.' One of his favorite sayings was, 'Work for the world.'

Marx made no secret of his teaching, it was open to all, even the capitalists, to read and understand. Nevertheless his offered prophecies were disregarded even as, one after the other, they were realized. The ruling class could not understand them—because they could not face the logical picture they revealed. And yet they were obliged to execute them, even to their own destruction.

In the course of the century since Marx's first analysis of capitalism, and largely through the utilization of science, productive methods have enormously improved in efficiency. Yet this great increase in productive power has not diminished in any degree the difficulties and contradictions of capitalism. In fact, as we all know from our bitter experience, it has very much increased them. From 1850 to 1950 we have witnessed crises growing in depth and duration, and resolving themselves in wars and reactionary tyrannies worse than anything that could have been imagined by any mid-nineteenth century economist or historian. But we have also witnessed the practical realization of Marx's more positive prophecies in the establishment of the first socialist state, the Soviet Union, which has been able, in spite of every opposition and attack, to grow and prosper. And there are now growing round it, west and east, other states imbued with the same creative philosophy. . . .

Bibliography

1. In *Capital*, Vol. I, Chap. XV.
2. See C. Babbage, *On the Economy of Machinery and Manufactures*, (London, 1832).

PAUL GOODMAN

Causerie at the Military-Industrial

Paul Goodman was a lecturer, teacher, critic, and novelist. He has written
extensively on society and education. In this article Goodman attacks the
tendency of the military-industrial complex to monopolize, absorb, and waste
vast technical and human resources for imperialistic ends. His remarks, often
polemical, confirm the warning and uneasiness about the relation of science to
war industry.

The National Security Industrial Association (NSIA) was founded in 1944 by
James Forrestal, to maintain and enhance the beautiful wartime communication
between the armament industries and the government. At present it comprises
400 members, including of course all the giant aircraft, electronics, motors, oil,
and chemical corporations, but also many one would not expect: not only
General Dynamics, General Motors, and General Telephone and Electronics, but
General Foods and General Learning; not only Sperry Rand, RCA, and
Lockheed, but Servco and Otis Elevators. It is a wealthy club. The military
budget is $84 billion.

At the recent biennial symposium, held on October 18 and 19 in the State
Department auditorium, the theme was "Research and Development in the
1970s." To my not unalloyed pleasure, I was invited to participate as one of the
seventeen speakers and assigned the topic "Planning for the Socio-Economic
Environment." Naturally I could make the usual speculations about why I was
thus "co-opted." I doubt that they expected to pick my brains for any profitable
ideas. But it is useful for feeders at the public trough to present an image of
wide-ranging discussion. It is comfortable to be able to say, "You see? these far-
outniks are impractical." And business meetings are dull and I am notoriously
stimulating. But the letter of invitation from Henri Busignies of ITT, the chairman of the symposium committee, said only, "Your accomplishments throughout
your distinguished career eminently qualify you to speak with authority on the
subject."

What is an intellectual man to do in such a case? I agree with the Gandhian
principle, always cooperate within the limits of honor, truth, and justice. But
how to cooperate with the military industrial club! during the Vietnam war
1967! It was certainly not the time to reason about basic premises, as is my
usual approach, so I decided simply to confront them and soberly tell them off.

Fortunately it was the week of the demonstration at the Pentagon, when there
would be thousands of my friends in Washington. So I tipped them off and
thirty students from Cornell and Harvard drove down early to picket the auditorium, with a good leaflet about the evil environment for youth produced by

the military corporations. When they came, the white helmets sprang up, plus the cameras and reporters. In the face of this dangerous invasion, the State Department of the United States was put under security, the doors were bolted, and the industrialists (and I) were not allowed to exit—on the 23rd Street side. Inside, I spoke as follows.

Research and Development for the Socio-Economic Environment of the 1970s

I am astonished that at a conference on planning for the future, you have not invited a single speaker under the age of thirty, the group that is going to live in that future. I am pleased that some of the young people have come to pound on the door anyway, but it is too bad that they aren't allowed to come in.

This is a bad forum for this topic. Your program mentions the "emerging national goals" of urban development, continuing education, and improving the quality of man's environment. I would add another essential goal, reviving American democracy; and at least two indispensable international goals, to rescue the majority of mankind from deepening poverty, and to insure the survival of mankind as a species. These goals indeed require research experimentation of the highest sophistication, but not by you. You people are unfitted by your commitments, your experience, your customary methods, your recruitment, and your moral disposition. You are the military industrial of the United States, the most dangerous body of men at the present in the world, for you not only implement our disastrous policies but are an overwhelming lobby for them, and you expand and rigidify the wrong use of brains, resources, and labor so that change becomes difficult. Most likely the trends you represent will be interrupted by a shambles of riots, alienation, ecological catastrophes, wars, and revolutions, so that current long-range planning, including this conference, is irrelevant. But if we ask what *are* the technological needs and what ought to be researched in this coming period, in the six areas I have mentioned, the best service that you people could perform is rather rapidly to phase yourselves out, passing on your relevant knowledge to people better qualified, or reorganizing yourselves with entirely different sponsors and commitments, so that you learn to think and feel in a different way. Since you are most of the R & D that there is, we cannot do without you as people, but we cannot do with you as you are.

In aiding technically underdeveloped regions, the need in the foreseeable future is for an intermediate technology, scientifically sophisticated but tailored to their local skills, tribal or other local social organization, plentiful labor force, and available raw materials. The aim is to help them out of starvation, disease, and drudgery without involving them in an international cash nexus of an entirely different order of magnitude. Let them take off at their own pace and in their own style. For models of appropriate technical analyses, I recommend you to E.F. Schumacher, of the British Coal Board, and his associates. Instead, you people—and your counterparts in Europe and Russia—have been imposing your technology, seducing native elites mostly corrupted by Western education, arming them, indeed often using them as a dumping ground for obsolete weapons. As Dr. Busignies pointed out yesterday, your aim must be, while

maintaining leadership, to allow very little technical gap, in order to do business.
Thus, you have involved these people in a wildly inflationary economy, have
driven them into instant urbanization, and increased the amount of disease and
destitution. You have disrupted ancient social patterns, debauched their cultures,
fomented tribal and other wars, and in Vietnam yourselves engaged in genocide.
You have systematically entangled them in Great Power struggles. It is not in
your interest, and you do not have the minds or the methods, to take these
peoples seriously as people.

The survival of the human species, at least in a civilized state, demands radical
disarmament, and there are several feasible political means to achieve this if we
willed it. By the same token, we must drastically de-energize the archaic system
of nation-states, e.g., by internationalizing space exploration, expanding operations like the International Geophysical Year, denationalizing Peace Corps and
aid programs, opening scientific information and travel. Instead, you—and your
counterparts in Europe, Russia, and China—have rigidified and aggrandized the
states with a Maginot-line kind of policy called Deterrence, which has continually escalated rather than stabilized. As Jerome Wiesner has demonstrated,
past a certain point your operations have increased insecurity rather than
diminished it. But this has been to your interest. Even in the present condition
of national rivalry, it has been estimated, by Marc Raskin who sat in on the
National Security Council, that the real needs of our defense should cost less
than a fourth of the budget you have pork-barreled. You tried, unsuccessfully,
to saddle us with the scientifically ludicrous Civil Defense program. You have
sabotaged the technology of inspection for disarmament. Now you are saddling
us with the antimissile missiles and the multi-warhead missiles (MIRV). You
have corrupted the human adventure of space with programs for armed plat-
forms in orbit. Although we are the most heavily armed and the most naturally
protected of the Great Power, you have seen to it that we spend a vastly greater
amount and perhaps a higher proportion of our wealth on armaments than any
other nation.
    This brings me to your effect on the climate of the economy. The wealth of a
nation is to provide useful goods and services, with an emphasis first on necessities and broad-spread comforts, simply as a decent background for uneconomic
life and culture; an indefinitely expanding economy is a rat-race. There ought
to be an even spread regionally, and no group must be allowed to fall outside of
society. At present, thanks to the scientific ingenuity and hard work of previous
generations, we could in America allow a modest livelihood to everyone as a
constitutional right. And on the other hand, as the young have been saying by
their style and actions, there is an imperative need to simplify the standard of
living, since the affluent standard has become frivolous, tawdry, and distracting
from life itself. But you people have distorted the structure of a rational
economy. Since 1945, half of new investment has gone into your products,
not subject to the market nor even to Congressional check. This year, 86
percent of money for research is for your arms and rockets. You push through
the colossally useless Super-Sonic Transport. At least 20 percent of the economy

is directly dependent on your enterprises. The profits and salaries of these enterprises are not normally distributed but go heavily to certain groups while others are excluded to the point of being outcaste. Your system is a major factor in producing the riots in Newark. [*At this remark there were indignant protests.*]

Some regions of the country are heavily favored—especially Pasadena and Dallas—and others disadvantaged. Public goods have been neglected. A disproportionate share of brains has been drained from more useful invention and development. And worst of all, you have enthusiastically supported an essentially mercantilist economics that measures economic health in terms of abstract Gross National Product and rate of growth, instead of concrete human well-being. Both domestically and internationally, you have been the bellwether of meaningless expansion, and this has sharpened poverty in our own slums and rural regions and for the majority of mankind. It has been argued that military expenditure, precisely because it is isolated and wasteful, is a stabilizer of an economy, providing employment and investment opportunities when necessary; but your unbridled expansion has been the chief factor of social instability.

Dramatically intervening in education, you have again disrupted the normal structure. Great universities have come to be financed largely for your programs. Faculties have become unbalanced; your kind of people do not fit into the community of scholars. The wandering dialogue of science with the unknown is straitjacketed for petty military projects. You speak increasingly of the need for personal creativity, but this is not to listen to the Creator Spirit for ideas, but to harness it to your ideas. This is blasphemous. There has been secrecy, which is intolerable to true academics and scientists. The political, and morally dubious co-opting of science, engineering, and social science has disgusted and alienated many of the best students. Further, you have warped the method of education, beginning with the primary grades. Your need for narrowly expert personnel has led to processing the young to be test-passers, with a gross exaggeration of credits and grading. You have used the wealth of public and parents to train apprentices for yourselves. Your electronics companies have gone into the "education industries" and tried to palm off teaching machines, audio-visual aids, and programmed lessons in excess of the evidence for their utility. But the educational requirements of our society in the foreseeable future demand a very different spirit and method. Rather than processing the young, the problem is how to help the young grow up free and inventive in a highly scientific and socially complicated world. We do not need professional personnel so much as autonomous professionals who can criticize the programs handed to them and be ethically responsible. Do you encourage criticism of your programs by either the subsidized professors or the students? [*At this, Mr. Charles Herzfeld, the chairman of the session, shouted "Yes!" and there was loud applause for the interruption, yet I doubt that there is much such encouragement.*] We need fewer lessons and tests, and there ought to be much less necessity and prestige attached to mandarin requirements.

Let us turn to urbanism. *Prima facie*, there are parts of urban planning—construction, depollution, the logistics of transportation—where your talents ought to be peculiarly useful. Unfortunately, it is your companies who have oversold the planes and the cars, polluted the air and water, and balked at even trivial remedies, so that I do not see how you can be morally trusted with the job. The chief present and future problems in this field, however, are of a different kind. They are two. The long-range problem is to diminish the urbanization and suburban sprawl altogether, for they are economically unviable and socially harmful. For this, the most direct means, and the one I favor, is to cut down rural emigration and encourage rural return, by means of rural reconstruction and regional cultural development. The aim should be a 20 percent rural ratio instead of the present 5 percent. This is an aspect of using high technology for simplification, increasing real goods but probably diminishing the Gross National Product measured in cash. Such a program is not for you. Your thinking is never to simplify and retrench, but always to devise new equipment to alleviate the mess that you have helped to make with your previous equipment.

   Secondly, the immediately urgent urban problem is how to diminish powerlessness, anomie, alienation, and mental disease. For this the best strategy is to decentralize urban administration, in policing, schooling, social welfare, neighborhood renewal, and real-estate and business ownership. Such community development often requires heightening conflict and risking technical inefficiency for intangible gains of initiative and solidarity. This also is obviously not your style. You want to concentrate capital and power. Your systems analyses of social problems always tend toward standardization, centralization, and bureaucratic control, although these are not necessary in the method. You do not like to feed your computers indefinite factors and unknown parameters where spirit, spite, enthusiasm, revenge, invention, etc., will make the difference. To be frank, your programs are usually grounded in puerile theories of social psychology, political science, and moral philosophy. There is a great need for research and trying out in this field, but the likely cast of characters might be small farmers, Negro matriarchs, political activists, long-haired students, and assorted sages. Not you. Let's face it. You are essentially producers of exquisite hardware and good at the logistics of moving objects around, but mostly with the crude aim of destroying things rather than reconstructing or creating anything, which is a harder task. Yet you boldly enter into fields like penology, pedagogy, hospital management, domestic architecture, and planning the next decade—wherever there is a likely budget.

In a society that is cluttered, overcentralized, and overadministered, we should aim at simplification, decentralization, and decontrol. These require highly sophisticated research to determine where, how, and how much. Further, for the first time in history, the scale of the artificial and technological has dwarfed the natural landscape. In prudence, we must begin to think of a principled limitation on artifice and to cut back on some of our present gigantic impositions, if only to insure that we do not commit some terrible ecological blunder. But as Dr. Smelt of Lockheed explained to us yesterday, it is the genius of American technology to go very rapidly from R & D to application; in this context, he

said, prudence is not a virtue. A particular case is automation: which human
functions should be computerized or automated, which should not? This
question—it is both an analytic and an empirical one—ought to be critical in the
next decade, but I would not trust IBM salesmen to solve it. Another problem
is how man can feel free and at home within the technological environment
itself. For instance, comprehending a machine and being able to repair it is one
thing; being a mere user and in bondage to service systems is another. Also, to
feel free, a man must have a rather strong say in the close environment that he
must deal with. But these requirements of a technology are not taken into
account by you. Despite Dr. Smelt, technology *is* a branch of moral philosophy,
subordinate to criteria like prudence, modesty, safety, amenity, flexibility,
cheapness, easy comprehension, repairability, and so forth. If such moral criteria
became paramount in the work of technologists, the quality of the environment
would be more livable.

Still a further problem is how to raise the scientific and technical culture of
the whole people, and here your imperialistic grab of the R & D money and of
the system of education has done immeasurable damage. You have seen to it
that the lion's share has gone to your few giant firms and a few giant universities,
although in fact very many, perhaps more than half of, important innovations
still come from independents and tiny firms. I was pleased that Dr. Dessauer of
Xerox pointed this out this morning. If the money were distributed more
widely, there would probably be more discovery and invention, and what is
more important, there would be a larger pool of scientific and competent
people. You make a fanfare about the spinoff of a few socially useful items,
but your whole enterprise is notoriously wasteful—for instance, five billions go
down the drain when after a couple of years you change the design of a sub-
marine, sorry about that. When you talk about spinoff, you people remind me
of the TV networks who, after twenty years of nothing, boast that they did
broadcast the McCarthy hearings and the Kennedy funeral. [*This remark led
to free and friendly laughter; I do not know whether at the other industry or
at their own hoax.*] Finally, concentrating the grants, you narrow the field of
discovery and innovation, creating an illusion of technological determinism, as if
we *had* to develop in a certain style. But if we had put our brains and money
into electric cars, we would now have electric cars; if we had concentrated on
intensive agriculture, we would now find that this is the most efficient, and so
forth. And in grabbing the funds, you are not even honest; 90 percent of the
R & D money goes in fact to shaping up for production, which as entrepreneurs
you should pay out of your own pockets.

No doubt some of these remarks have been unfair and ignorant. [*Frantic
applause.*] By and large they are undeniable, and I have not been picking nits.

These remarks have certainly been harsh and moralistic. We are none of us
saints, and ordinarily I would be ashamed to use such a tone. But you are the
manufacturers of napalm, fragmentation bombs, the planes that destroy rice.
Your weapons have killed hundreds of thousands in Vietnam and you will kill
other hundreds of thousands in other Vietnams. I am sure that most of you
would concede that much of what you do is ugly and harmful, at home and

abroad. But you would say that it is necessary for the American way of life, at home and abroad, and therefore you cannot do otherwise. Since we believe, however, that that way of life itself is unnecessary, ugly, and un-American [*Shouts of "Who are we?"*] —we are I and those people outside—we cannot condone your present operations; they should be wiped off the slate.

WALTER WEISSKOPF

Alienation and Economics

Walter A. Weisskopf is Professor of Economics at Roosevelt University in Illinois. He made important contributions to the philosophy of economics and his book, *The Psychology of Economics* (1955), is considered the definitive text in its field. In this selection, Weisskopf critically analyzes the concepts of economic growth and consumption in relation to expanding technology. He concludes that technological and economic expansion, while creating and satisfying superfluous needs, fails to meet more fundamental human needs of equilibrium and stability. Modern, technological economies, he argues, generate insecurity and alienation instead of meeting collective human requirements. Economic growth is historically obsolete.

Economic Growth and Need Satisfaction

. . . The conflict between the concept of economic growth and the concept of an equilibrium of satisfaction reflects an inner conflict in modern capitalism. As we have seen, its last phase, the corporate state, elevated economic growth together with technological innovation to the dignity of an ideal. At the same time, economics, the self-interpretation of this society, maintained that the economy, including the giant corporations, serve the satisfaction of human needs. If such satisfaction depends on the achievement of an equilibrium, of a balance, of satiation and contentment, the corporate and national striving for more and more would, at some point, become meaningless. Therefore, econo mists have tried to combine the two incompatible concepts by trying to establish a *need* for unlimited economic growth, and thus give it a meaning and justification. This was done by trying to identify needs in general with "natural" physiological needs and, in turn, to present the idea of continuous economic growth as rooted in nature.

The reason for this approach can be found in the naturalistic, scientistic style of economic thought and in its ideological implications. If something can be explained as "natural," as rooted in biological instincts, drives or needs, it is, in a scientific culture, implicitly justified. Implicitly economists assume that

needs are given by nature, originating with the individual, and that production
and the economic system serve to satisfy those needs.

This situation, however, is by no means universal; it is abstracted from
primitive economics where mere subsistence is the goal of economic action. It
existed in all economies where agriculture is the main economic activity and
it exists in some of the most underdeveloped countries today. In such situations
the goal of economic activity is survival. Human needs here are, indeed, "given"
by biological necessity. Here, man is subject to the dire necessity of working
and economizing. Only these activities ensure his survival. Here, economic
activity is part of the metabolic cycle of life; production, consumption, digestion,
fertilization and renewed production follow each other in a never-ending cycle.
They are subject to the laws of nature. As far as economic activity was unavoidable through the pressure of necessity, it did not require any formulation of
goals for its justification or meaning. It is in the nature of an automatic movement, such as the beating of the heart. Only insofar as economic activity
produces a surplus over and above the bare necessities for survival does it enter
the realm of freedom and require meaningful justification. This meaning was,
in preindustrial society, created by religious, social and political institutions in
which the economic activity was embedded. The surplus was used for the
purposes of groups which were the carriers of the world outlook and meaning
of their societies. The surplus over and above the means of subsistence and the
economic activity producing it were considered as means to noneconomic ends
and received their meaning from religious, social and political institutions.
This may very well be the only meaningful evaluation of economic activity; as
far as it is not subject to necessity, it can only receive a meaning from something
that is beyond the economic dimension. Otherwise, economic activity becomes
meaningless.

In order to avoid this result and in the absence of a superordinated system of
noneconomic values in Western society, economic thought tried to reduce all
economic activity to the level of physiological need satisfaction. If everything
we are producing, selling and consuming today in the advanced complex
Western economies could be interpreted as satiating a quasi-physiological drive
or need, our economic activity would need no further justification.

The physiological interpretation of need satisfaction presupposes a state of
tension which is relieved by satisfaction. Without the emergence of tension,
no satisfaction can be received from physiological need satisfaction. Without
hunger the intake of food is not pleasurable. This statement is so trite that one
hesitates to make it. However, this obvious truth is overlooked when it comes
to an evaluation of economic acquisition and growth. The principle of declining
marginal utility can be applied not only to specific kinds of needs and their
satisfaction but also to the entire field of acquisition of goods and services
which are produced by our economy.

The affluent industrial economy has immensely enlarged the field of "need
satisfaction" and raised so-called standards of living to a peak that was never
reached before. We have come such a long way that many social scientists and
philosophers talk about an imminent state of affairs in which men will be

freed of economic necessity altogether by automation and cybernation of production. They predict that, in the foreseeable future, man in the advanced industrial economy may be able to make the "leap from the realm of necessity into the realm of freedom." However that may be, even at the present time, the advanced Western economies have reached a level of production which has reduced immensely the importance of acquiring more and more additional goods and services of the traditional variety. This is not a situation of which people are conscious. The ideology of "more and more" is still so strong that people are precluded from becoming aware of the fact that they are forced into more work and more acquisition by the socioeconomic system rather than by their free inclination. Intensive advertising and the all-prevasive fact of artificial obsolescence are clear and present symptoms of this situation. Artificial obsolescence through planned change is the social correlate to the emergence of tension in the physiological field. *Nolens volens*, people get hungry. Planned obsolescence replaces physiological tension in those fields where no natural automatic tension arises. What firms and advertisers are doing is to create hunger where nature has not provided for such an incidence. Through frequently changing styles of cars and clothes, and by exploiting the desire for conformity, they "force" people to develop a "need" for change much like the need for food that arises automatically for physiological reasons. The same purpose is accomplished by the continuous development of new products and new kinds of need satisfaction. Once the new product and the new kind of "need" is marketed, the pressure of conformity actually creates a "need." This is the way in which business practice and economic thought try to combine a restrictive natural physiological interpretation of needs with the idea of economic growth. The physiological pattern of recurrent tension is artificially reproduced, thereby representing economic growth as the result of "natural" physiological forces. That purely physical and sensual pleasure requires ever more excitation and titillation, tension and pain, was known not only to the Hindus and Buddhists but also to the Greek philosophers of the Periclean and Hellenistic periods. It was, of course, known to Christian thought from the Fathers of the Church to the Middle Ages. All these cultures saw virtue in balance and moderation, or even in restraint and negation of desires. Modern civilization, however, has elevated the continuous creation of tension for the sake of pleasurable "satisfaction" to the dignity of an ultimate goal. Economic growth as an individual and social ideal seems justified if this constant creation of new "needs" can be viewed as a continuation of a natural process. However, what is overlooked is that in nature a point of satiation is reached, whereas, in the affluent economy, this is prevented by the continuous creation of new "needs. . . ."

Economic and technical change are inseparable from economic growth. It is highly dubious that the magnitude and pace of change in the Western economies are conducive to the satisfaction of human needs; on the contrary they continuously prevent the establishment of a satisfactory equilibrium.

There is a basic need for a balance between growth, change, dynamics on the one hand, and stability, constancy, rest, security on the other hand. This balance is disturbed by the "gale of creative destruction" in industrial society. This gale and its destruction have made life in the Western economies extremely unstable. Erik Erikson once said that assimilation to the American scene involves adaptation to constant change, in itself a difficult and burdensome way of life. Kenneth Keniston considers chronic technological and economic change as one of the major alienating elements in American society. The continuous change inspired by science, technology and business certainly makes life and the future more insecure than ever. Life planning becomes almost impossible. Chosen careers and learned skills become obsolete. Even the most primitive skills such as writing and arithmetic are obliterated by typewriters and computers. Even the physical environment changes at breakneck speed. Cities and buildings are destroyed and new ones are erected within weeks. What was there yesterday is not there today. An inhabitant of New York or Chicago who was abroad for a year may have difficulties in finding his way around in the city. A psychoanalyst once defined the hometown as the place where one always knows what is around the corner. In that sense the modern urbanite has no home.

An increasingly faster rate of change is characteristic of our present economy. Although much attention is paid to the difficulties of adjustment to a rapidly changing economy, there is hardly any discussion of an optimal rate of change. Is there a limit to the degree of economic change to which man can adjust? Is there a limit beyond which economic change, even if desirable in terms of more output, becomes intolerable for individuals and detrimental to well-being? It is, of course, difficult to give an answer to this question in quantitative terms; but it should be established that change is not always a gain but can inflict psychological discomfort and suffering.

The fact that statistical measurement of these psychological costs is difficult or impossible is no excuse for ignoring them. After all, economic thought made its greatest strides long before it became mathematical and statistical. The great economists from Adam Smith to Keynes based many of their findings on introspection. Insight can be used in detecting the hidden costs of economic progress. Bertrand de Jouvenel suggests the use of sensitive persons to detect the impending destructive effects of new types of production; they could be used like radar or a Geiger counter in such cases.

Economic growth with its gale of destruction has contributed to the alienation, insecurity and rootlessness of Western man. If he has found roots in a stable environment, economic change may uproot him again. The general feeling of insecurity and lack of community that pervades our society may stem from continuous change and its threat to physical and mental stability. The modern economy forces man into a pattern of extreme flexibility and detachment. He has to be continuously on the alert and adjust himself to the changing frontiers of production, jobs and consumption. This has made him into a lonely member of a crowd. Economic change may sever the ties of habitat and neighborhood; it may cut apart the bonds of friendship and human relations. The great attention paid to human relations in industry is a consequence of the lack of attachment and involvement that continuous change requires. Definitely there can be too much change and too fast a rate of change. Human beings require an equilibrium between change and stability, a need which is not met by the modern economy. . . .

C. P. SNOW

Two Cultures: And A Second Look

C.P. Snow is an English novelist and scientist. He is also a Fellow of Christ College, Cambridge University, England. Snow has been active in government and has published a series of novels under the general title "Strangers and Brothers." He has written and lectured extensively on the social relations of science. Snow's essay exposes the rift, tension, and misunderstanding between persons trained in the sciences and persons trained in humanistic disciplines. His contention is that humanists, for the most part, and literary intellectuals in particular, have refused to integrate the scientific culture into their understanding. In their passionate anti-scientific attitudes they encourage an aimless science, and propagate misconceptions about science. This has had the effect of inhibiting the full realization of science as a human and humane institution. Before it is too late, humanist intellectuals must recognize that the future belongs to science or there will be no future at all. Men of learning have an obligation to understand science and contribute to its growth and control for human and social purposes.

It is about three years since I made a sketch in print of a problem which had been on my mind for some time. It was a problem I could not avoid just because of the circumstances of my life. The only credentials I had to ruminate on the subject at all came through those circumstances, through nothing more than a set of chances. Anyone with similar experience would have seen much the same things and I think made very much the same comments about them. It just happened to be an unusual experience. By training I was a scientist: by vocation I was a writer. That was all. It was a piece of luck, if you like, that arose through coming from a poor home.

But my personal history isn't the point now. All that I need say is that I came to Cambridge and did a bit of research here at a time of major scientific activity. I was privileged to have a ringside view of one of the most wonderful creative periods in all physics. And it happened through the flukes of war—including meeting W.L. Bragg in the buffet on Kettering station on a very cold morning in 1939, which had a determining influence on my practical life—that I was

Reprinted with permission from Cambridge University Press. From *Two Cultures: And A Second Look* by C.P. Snow. New York: Cambridge University Press, 1969, pp. 1–29.

able, and indeed morally forced, to keep that ringside view ever since. So for thirty years I have had to be in touch with scientists not only out of curiosity, but as part of a working existence. During the same thirty years I was trying to shape the books I wanted to write, which in due course took me among writers.

There have been plenty of days when I have spent the working hours with scientists and then gone off at night with some literary colleagues. I mean that literally. I have had, of course, intimate friends among both scientists and writers. It was through living among these groups and much more, I think, through moving regularly from one to the other and back again that I got occupied with the problem of what, long before I put it on paper, I christened to myself as the 'two cultures'. For constantly I felt I was moving among two groups—comparable in intelligence, identical in race, not grossly different in social origin, earning about the same incomes, who had almost ceased to communicate at all, who in intellectual, moral and psychological climate had so little in common that instead of going from Burlington House or South Kensington to Chelsea, one might have crossed an ocean.

In fact, one had travelled much further than across an ocean—because after a few thousand Atlantic miles, one found Greenwich Village talking precisely the same language as Chelsea, and both having about as much communication with M.I.T. as though the scientists spoke nothing but Tibetan. For this is not just our problem; owing to some of our educational and social idiosyncrasies, it is slightly exaggerated here, owing to another English social peculiarity it is slightly minimized; by and large this is a problem of the entire West.

By this I intend something serious. I am not thinking of the pleasant story of how one of the more convivial Oxford greats dons—I have heard the story attributed to A.L. Smith—came over to Cambridge to dine. The date is perhaps the 1890's. I think it must have been at St. John's, or possibly Trinity. Anyway, Smith was sitting at the right hand of the President—or Vice-Master—and he was a man who liked to include all round him in the conversation, although he was not immediately encouraged by the expressions of his neighbors. He addressed some cheerful Oxonian chit-chat at the one opposite to him, and got a grunt. He then tried the man on his own right hand and got another grunt. Then, rather to his surprise, one looked at the other and said, 'Do you know what he's talking about?' 'I haven't the least idea.' At this, even Smith was getting out of his depth. But the President, acting as a social emollient, put him at his ease, by saying, 'Oh, those are mathematicians! We never talk to *them*'.

No, I intend something serious. I believe the intellectual life of the whole of Western society is increasingly being split into two polar groups. When I say the intellectual life, I mean to include also a large part of our practical life, because I should be the last person to suggest the two can at the deepest level be distinguished. I shall come back to the practical life a little later. Two polar groups: at one pole we have the literary intellectuals, who incidentally while no one was looking took to referring to themselves as 'intellectuals' as though there were no others. I remember G.H. Hardy once remarking to me in mild puzzlement, some time in the 1930's: 'Have you noticed how the word "intellectual" is used nowadays? There seems to be a new definition which certainly doesn't include Rutherford or Eddington or Dirac or Adrian or me. It does seem rather odd, don't y'know.'

Literary intellectuals at one pole—at the other scientists, and as the most representative, the physical scientists. Between the two a gulf of mutual incomprehension—sometimes (particularly among the young) hostility and dislike, but most of all lack of understanding. They have a curious distorted image of each other. Their attitudes are so different that, even on the level of emotion, they can't find much common ground. Nonscientists tend to think of scientists as brash and boastful. They hear Mr. T.S. Eliot, who just for these illustrations we can take as an archetypal figure, saying about his attempts to revive verse-drama, that we can hope for very little, but that he would feel content if he and his coworkers could prepare the ground for a new Kyd or a new Greene. That is the tone, restricted and constrained, with which literary intellectuals are at home: it is the subdued voice of their culture. Then they hear a much louder voice, that of another archetypal figure, Rutherford, trumpeting: 'This is the heroic age of science! This is the Elizabethan age!' Many of us heard that, and a good many other statements beside which that was mild; and we weren't left in any doubt whom Rutherford was casting for the role of Shakespeare. What is hard for the literary intellectuals to understand, imaginatively or intellectually, is that he was absolutely right.

And compare 'this is the way the world ends, not with a bang but a whimper'—incidentally, one of the least likely scientific prophecies ever made—compare that with Rutherford's famous repartee, 'Lucky fellow, Rutherford, always on the crest of the wave.' 'Well, I made the wave, didn't I?'

The nonscientists have a rooted impression that the scientists are shallowly optimistic, unaware of man's condition. On the other hand, the scientists believe that the literary intellectuals are totally lacking in foresight, peculiarly unconcerned with their brother men, in a deep sense anti-intellectual, anxious to restrict both art and thought to the existential moment. And so on. Anyone with a mild talent for invective could produce plenty of this kind of subterranean back-chat. On each side there is some of it which is not entirely baseless. It is all destructive. Much of it rests on misinterpretations which are dangerous. I should like to deal with two of the most profound of these now, one on each side.

First, about the scientists' optimism. This is an accusation which has been made so often that it has become a platitude. It has been made by some of the acutest nonscientific minds of the day. But it depends upon a confusion between the individual experience and the social experience, between the individual condition of man and his social condition. Most of the scientists I have known well have felt—just as deeply as the nonscientists I have known well—that the individual condition of each of us is tragic. Each of us is alone: sometimes we escape from solitariness, through love or affection or perhaps creative moments, but those triumphs of life are pools of light we make for ourselves while the edge of the road is black: each of us dies alone. Some scientists I have known have had faith in revealed religion. Perhaps with them the sense of the tragic condition is not so strong. I don't know. With most people of deep feeling, however high-spirited and happy they are, sometimes most with those who are happiest and most high-spirited, it seems to be right in the fibers, part of the weight of life. That is as true of the scientists I have known best as of anyone at all.

But nearly all of them—and this is where the color of hope genuinely comes in—would see no reason why, just because the individual condition is tragic, so must the social condition be. Each of us is solitary: each of us dies alone: all right, that's a fate against which we can't struggle—but there is plenty in our condition which is not fate, and against which we are less than human unless we do struggle.

Most of our fellow human beings, for instance, are underfed and die before their time. In the crudest terms, *that* is the social condition. There is a moral trap which comes through the insight into man's loneliness: it tempts one to sit back, complacent in one's unique tragedy, and let the others go without a meal.

As a group, the scientists fall into that trap less than others. They are inclined to be impatient to see if something can be done: and inclined to think that it can be done, until it's proved otherwise. That is their real optimism, and it's an optimism that the rest of us badly need.

In reverse, the same spirit, tough and good and determined to fight it out at the side of their brother men, has made scientists regard the other culture's social attitudes as contemptible. That is too facile: some of them are, but they are a temporary phase and not to be taken as representative.

I remember being cross-examined by a scientist of distinction. 'Why do most writers take on social opinions which would have been thought distinctly uncivilized and démodé at the time of the Plantagenets? Wasn't that true of most of the famous twentieth-century writers? Yeats, Pound, Wyndham Lewis, nine out of ten of those who have dominated literary sensibility in our time—weren't they not only politically silly, but politically wicked? Didn't the influence of all they represent bring Auschwitz that much nearer?'

I thought at the time, and I still think, that the correct answer was not to defend the indefensible. It was no use saying that Yeats, according to friends whose judgment I trust, was a man of singular magnanimity of character, as well as a great poet. It was no use denying the facts, which are broadly true. The honest answer was that there is, in fact, a connection, which literary persons were culpably slow to see, between some kinds of early twentieth-century art and the most imbecile expressions of antisocial feeling. That was one reason, among many, why some of us turned our backs on the art and tried to hack out a new or different way for ourselves.

But though many of those writers dominated literary sensibility for a generation, that is no longer so, or at least to nothing like the same extent. Literature changes more slowly than science. It hasn't the same automatic corrective, and so its misguided periods are longer. But it is ill-considered of scientists to judge writers on the evidence of the period 1914–50.

Those are two of the misunderstandings between the two cultures. I should say, since I began to talk about them—the two cultures, that is—I have had some criticism. Most of my scientific acquaintances think that there is something in it, and so do most of the practicing artists I know. But I have been argued with by nonscientists of strong down-to-earth interests. Their view is that it is an oversimplification, and that if one is going to talk in these terms there ought to be at least three cultures. They argue that, though they are not scientists themselves, they would share a good deal of the scientific feeling. They would have as little use—perhaps, since they knew more about it, even less use—for the recent literary culture as the scientists themselves. J.H. Plumb, Alan Bullock and some of my American sociological friends have said that they vigorously refuse to be corralled in a cultural box with people they wouldn't be seen dead with, or to be regarded as helping to produce a climate which would not permit of social hope.

I respect those arguments. The number 2 is a very dangerous number: that is why the dialectic is a dangerous process. Attempts to divide anything into two ought to be regarded with much suspicion. I have thought a long time about going in for further refinements: but in the end I have decided against. I was searching for something a little more than a dashing metaphor, a good deal less than a cultural map: and for those purposes the two cultures is about right, and subtilizing any more would bring more disadvantages than it's worth.

At one pole, the scientific culture really is a culture, not only in an intellectual but also in an anthropological sense. That is, its members need not, and of course often do not, always completely understand each other; biologists more often than not will have a pretty hazy idea of contemporary physics; but there are common attitudes, common standards and patterns of behavior, common approaches and assumptions. This goes surprisingly wide and deep. It cuts across other mental patterns, such as those of religion or politics or class.

Statistically, I suppose slightly more scientists are in religious terms unbelievers, compared with the rest of the intellectual world—though there are plenty who are religious, and that seems to be increasingly so among the young. Statistically also, slightly more scientists are on the Left in open politics—though again, plenty always have called themselves conservatives, and that also seems to be more common among the young. Compared with the rest of the intellectual world, considerably more scientists in this country and probably in the U.S. come from poor families. Yet, over a whole range of thought and behavior, none of that matters very much. In their working, and in much of their emotional life, their attitudes are closer to other scientists than to nonscientists who in religion or politics or class have the same labels as themselves. If I were to risk a piece of shorthand, I should say that naturally they had the future in their bones.

They may or may not like it, but they have it. That was as true of the conservatives J.J. Thomson and Lindemann as of the radicals Einstein or Blackett: as true of the Christian A.H. Compton as of the materialist Bernal: of the aristocrats Broglie or Russell as of the proletarian Faraday: of those born rich, like Thomas Merton or Victor Rothschild, as of Rutherford, who was the son of an odd-job handyman. Without thinking about it, they respond alike. That is what a culture means.

At the other pole, the spread of attitudes is wider. It is obvious that between the two, as one moves through intellectual society from the physicists to the literary intellectuals, there are all kinds of tones of feeling on the way. But I believe the pole of total incomprehension of science radiates its influence on all the rest. That total incomprehension gives, much more pervasively than we realize, living in it, an unscientific flavor to the whole 'traditional' culture, and that unscientific flavor is often, much more than we admit, on the point of turning anti-scientific. The feelings of one pole become the anti-feelings of the other. If the scientists have the future in their bones, then the traditional culture responds by wishing the future did not exist. It is the traditional culture, to an extent remarkably little diminished by the emergence of the scientific one, which manages the western world.

This polarization is sheer loss to us all. To us as people, and to our society. It is at the same time practical and intellectual and creative loss, and I repeat that it is false to imagine that those three considerations are clearly separable. But for a moment I want to concentrate on the intellectual loss.

The degree of incomprehension on both sides is the kind of joke which has gone sour. There are about fifty thousand working scientists in the country and about eighty thousand professional engineers or applied scientists. During the war and in the years since, my colleagues and I have had to interview somewhere between thirty to forty thousand of these—that is, about 25 per cent. The number is large enough to give us a fair sample, though of the men we talked to most would still be under forty. We were able to find out a certain amount of what they read and thought about. I confess that even I, who am fond of them and respect them, was a bit shaken. We hadn't quite expected that the links with the traditional culture should be so tenuous, nothing more than a formal touch of the cap.

As one would expect, some of the very best scientists had and have plenty of energy and interest to spare, and we came across several who had read everything that literary people talk about. But that's very rare. Most of the rest, when one tried to probe for what books they had read, would modestly confess, 'Well, I've *tried* a bit of Dickens,' rather as though Dickens were an extraordinarily esoteric, tangled and dubiously rewarding writer, something like Rainer Maria Rilke. In fact that is exactly how they do regard him: we thought that discovery, that Dickens had been transformed into the type-specimen of literary incomprehensibility, was one of the oddest results of the whole exercise.

But of course, in reading him, in reading almost any writer whom we should value, they are just touching their caps to the traditional culture. They have their own culture, intensive, rigorous, and constantly in action. This culture contains a great deal of argument, usually much more rigorous, and almost always at a higher conceptual level, than literary persons' arguments—even though the scientists do cheerfully use words in senses which literary persons don't recognize, the senses are exact ones, and when they talk about 'subjective,' 'objective,' 'philosophy' or 'progressive,' they know what they mean, even though it isn't what one is accustomed to expect.

Remember, these are very intelligent men. Their culture is in many ways an exacting and admirable one. It doesn't contain much art, with the exception, an important exception, of music. Verbal exchange, insistent argument. Long-playing records. Color-photography. The ear, to some extent the eye. Books, very little, though perhaps not many would go so far as one hero, who perhaps I should admit was further down the scientific ladder than the people I've been talking about—who, when asked what books he read, replied firmly and confidently: 'Books? I prefer to use my books as tools.' It was very hard not to let the mind wander—what sort of tool would a book make? Perhaps a hammer? A primitive digging instrument?

Of books, though, very little. And of the books which to most literary persons are bread and butter, novels, history, poetry, plays, almost nothing at all. It isn't that they're not interested in the psychological or moral or social life. In the social life, they certainly are, more than most of us. In the moral, they are by and large the soundest group of intellectuals we have; there is a moral component right in the grain of science itself, and almost all scientists form their own judgments of the moral life. In the psychological they have as much interest as most of us, though occasionally I fancy they come to it rather late. It isn't that they lack the interests. It is much more that the whole literature of the traditional culture doesn't seem to them relevant to those interests. They are, of course, dead wrong. As a result, their imaginative understanding is less than it could be. They are self-impoverished.

But what about the other side? They are impoverished too—perhaps more seriously, because they are vainer about it. They still like to pretend that the traditional culture is the whole of 'culture,' as though the natural order didn't exist. As though the exploration of the natural order was of no interest either in its own value or its consequences. As though the scientific edifice of the physical world was not, in its intellectual depth, complexity and articulation, the most beautiful and wonderful collective work of the mind of man. Yet most nonscientists have no conception of that edifice at all. Even if they want to have it, they can't. It is rather as though, over an immense range of intellectual experience, a whole group was tone-deaf. Except that this tone-deafness doesn't come by nature, but by training, or rather the absence of training.

As with the tone-deaf, they don't know what they miss. They give a pitying chuckle at the news of scientists who have never read a major work of English literature. They dismiss them as ignorant specialists. Yet their own ignorance and their own specialization is just as startling. A good many times I have been present at gatherings of people who, by the standards of the traditional culture, are thought highly educated and who have with considerable gusto been express-ing their incredulity at the illiteracy of scientists. Once or twice I have been provoked and have asked the company how many of them could describe the Second Law of Thermodynamics. The response was cold: it was also negative. Yet I was asking something which is about the scientific equivalent of: *Have you read a work of Shakespeare's?*

I now believe that if I had asked an even simpler question—such as, What do you mean by mass, or acceleration, which is the scientific equivalent of saying, *Can you read?*—not more than one in ten of the highly educated would have felt that I was speaking the same language. So the great edifice of modern physics goes up, and the majority of the cleverest people in the western world have about as much insight into it as their neolithic ancestors would have had.

Just one more of those questions, that my nonscientific friends regard as being in the worst of taste. Cambridge is a university where scientists and non-scientists meet every night at dinner. About two years ago, one of the most astonishing experiments in the whole history of science was brought off. I don't mean the sputnik—that was admirable for quite different reasons, as a feat of organization and a triumphant use of existing knowledge. No, I mean the experiment at Columbia by Yang and Lee. It is an experiment of the greatest beauty and originality, but the result is so startling that one forgets how beautiful the experiment is. It makes us think again about some of the fundamentals of the physical world. Intuition, common sense—they are neatly stood on their heads. The result is usually known as the contradiction of parity. If there were any serious communication between the two cultures, this experiment would have been talked about at every High Table in Cambridge. Was it? I wasn't here: but I should like to ask the question.

There seems then to be no place where the cultures meet. I am not going to waste time saying that this is a pity. It is much worse than that. Soon I shall come to some practical consequences. But at the heart of thought and creation we are letting some of our best chances go by default. The clashing point of two subjects, two disciplines, two cultures—of two galaxies, so far as that goes—ought to produce creative chances. In the history of mental activity that has been where some of the break-throughs came. The chances are there now. But they are there, as it were, in a vacuum, because those in the two cultures can't talk to each other. It is bizarre how very little of twentieth-century science has been assimilated into twentieth-century art. Now and then one used to find poets conscientiously using scientific expressions, and getting them wrong—there was a time when 'refraction' kept cropping up in verse in a mystifying fashion, and when 'polarized light' was used as though writers were under the illusion that it was a specially admirable kind of light.

Of course, that isn't the way that science could be any good to art. It has got to be assimilated along with, and as part and parcel of, the whole of our mental experience, and used as naturally as the rest.

I said earlier that this cultural divide is not just an English phenomenon: it exists all over the western world. But it probably seems at its sharpest in England, for two reasons. One is our fanatical belief in educational specialization, which is much more deeply ingrained in us than in any country in the world, West or East. The other is our tendency to let our social forms crystallize. This tendency appears to get stronger, not weaker, the more we iron out economic inequalities: and this is specially true in education. It means that once anything like a cultural divide gets established, all the social forces operate to make it not less rigid, but more so.

The two cultures were already dangerously separate sixty years ago; but a prime minister like Lord Salisbury could have his own laboratory at Hatfield, and Arthur Balfour had a somewhat more than amateur interest in natural science. John Anderson did some research in organic chemistry in Würzburg before passing first into the Civil Service, and incidentally took a spread of subjects which is now impossible. None of that degree of interchange at the top of the Establishment is likely, or indeed thinkable, now.

In fact, the separation between the scientists and nonscientists is much less bridgeable among the young than it was even thirty years ago. Thirty years ago the cultures had long ceased to speak to each other: but at least they managed a kind of frozen smile across the gulf. Now the politeness has gone, and they just make faces. It is not only that the young scientists now feel that they are part of a culture on the rise while the other is in retreat. It is also, to be brutal, that the young scientists know that with an indifferent degree they'll get a comfortable job, while their contemporaries and counterparts in English or History will be lucky to earn 60 per cent as much. No young scientist of any talent would feel that he isn't wanted or that his work is ridiculous, as did the hero of *Lucky Jim*, and in fact, some of the disgruntlement of Amis and his associates is the disgruntlement of the under-employed arts graduate.

There is only one way out of all this: it is, of course, by rethinking our education. In this country, for the two reasons I have given, that is more difficult than in any other. Nearly everyone will agree that our school education is too specialized. But nearly everyone feels that it is outside the will of man to alter it. Other countries are as dissatisfied with their education as we are, but are not so resigned.

The U.S. teach out of proportion more children up to eighteen than we do: they teach them far more widely, but nothing like so rigorously. They know that: they are hoping to take the problem in hand within ten years, though they may not have all that time to spare. The U.S.S.R. also teach out of proportion more children than we do: they also teach far more widely than we do (it is an absurd Western myth that their school education is specialized) but much too rigorously. They know that—and they are beating about to get it right. The Scandinavians, in particular the Swedes, who would make a more sensible job of it than any of us, are handicapped by their practical need to devote an inordinate amount of time to foreign languages. But they too are seized of the problem.

Are we? Have we crystallized so far that we are no longer flexible at all?

Talk to schoolmasters, and they say that our intense specialization, like nothing else on earth, is dictated by the Oxford and Cambridge scholarship examinations. If that is so, one would have thought it not utterly impracticable to change the Oxford and Cambridge scholarship examinations. Yet one would underestimate the national capacity for the intricate defensive to believe that that was easy. All the lessons of our educational history suggest we are only capable of increasing specialization, not decreasing it.

Somehow we have set ourselves the task of producing a tiny *élite*—far smaller proportionately than in any comparable country—educated in one academic skill. For a hundred and fifty years in Cambridge it was mathematics: then it was mathematics or classics: then natural science was allowed in. But still the choice had to be a single one.

It may well be that this process has gone too far to be reversible. I have given reasons why I think it is a disastrous process, for the purpose of a living culture. I am going on to give reasons why I think it is fatal, if we're to perform our practical tasks in the world. But I can think of only one example, in the whole of English educational history, where our pursuit of specialized mental exercises was resisted with success.

It was done here in Cambridge, fifty years ago, when the old order-of-merit in the Mathematical Tripos was abolished. For over a hundred years, the nature of the Tripos had been crystallizing. The competition for the top places had got fiercer, and careers hung on them. In most colleges, certainly in my own, if one managed to come out as Senior or Second Wrangler, one was elected a Fellow out of hand. A whole apparatus of coaching had grown up. Men of the quality of Hardy, Littlewood, Russell, Eddington, Jeans, Keynes, went in for two or three years' training for an examination which was intensely competitive and intensely difficult. Most people in Cambridge were very proud of it, with a similar pride to that which almost anyone in England always has for our existing educational institutions, whatever they happen to be. If you study the fly-sheets of the time, you will find the passionate arguments for keeping the examination precisely as it was to all eternity: it was the only way to keep up standards, it was the only fair test of merit, indeed, the only seriously objective test in the world. The arguments, in fact, were almost exactly those which are used today with precisely the same passionate sincerity if anyone suggests that the scholarship examinations might conceivably not be immune from change.

In every respect but one, in fact, the old Mathematical Tripos seemed perfect. The one exception, however, appeared to some to be rather important. It was simply—so the young creative mathematicians, such as Hardy and Littlewood, kept saying—that the training had no intellectual merit at all. They went a little further, and said that the Tripos had killed serious mathematics in England stone dead for a hundred years. Well, even in academic controversy, that took some skirting round, and they got their way. But I have an impression that Cambridge was a good deal more flexible between 1850 and 1914 than it has been in our time. If we had had the old Mathematical Tripos firmly planted among us, should we have ever managed to abolish it?

The reasons for the existence of the two cultures are many, deep, and complex, some rooted in social histories, some in personal histories, and some in the inner dynamic of the different kinds of mental activity themselves. But I want to isolate one which is not so much a reason as a correlative, something which winds in and out of any of these discussions. It can be said simply, and it is this. If we forget the scientific culture, then the rest of Western intellectuals have never tried, wanted, or been able to understand the industrial revolution, much less accept it. Intellectuals, in particular literary intellectuals, are natural Luddites.

That is specially true of this country, where the industrial revolution happened to us earlier than elsewhere, during a long spell of absentmindedness. Perhaps that helps explain our present degree of crystallization. But, with a little qualification, it is also true, and surprisingly true, of the United States.

In both countries, and indeed all over the West, the first wave of the industrial revolution crept on, without anyone noticing what was happening. It was, of course—or at least it was destined to become, under our own eyes, and in our own time—by far the biggest transformation in society since the discovery of agriculture. In fact, those two revolutions, the agricultural and the industrial-scientific, are the only qualitative changes in social living that men have ever known. But the traditional culture didn't notice: or when it did notice, didn't like what it saw. Not that the traditional culture wasn't doing extremely well out of the revolution; the English educational institutions took their slice of the English nineteenth-century wealth, and perversely, it helped crystallize them in the forms we know.

Almost none of the talent, almost none of the imaginative energy, went back into the revolution which was producing the wealth. The traditional culture became more abstracted from it as it became more wealthy, trained its young men for administration, for the Indian Empire, for the purpose of perpetuating the culture itself, but never in any circumstances to equip them to understand the revolution or take part in it. Farsighted men were beginning to see, before the middle of the nineteenth century, that in order to go on producing wealth, the country needed to train some of its bright minds in science, particularly in applied science. No one listened. The traditional culture didn't listen at all: and the pure scientists, such as there were, didn't listen very eagerly. You will find the story, which in spirit continues down to the present day, in Eric Ashby's *Technology and the Academics*.

The academics had nothing to do with the industrial revolution; as Corrie, the old Master of Jesus, said about trains running into Cambridge on Sunday, 'It is equally displeasing to God and to myself.' So far as there was any thinking in nineteenth-century industry, it was left to cranks and clever workmen. American social historians have told me that much the same was true of the U.S. The industrial revolution, which began developing in New England fifty years or so later than ours, apparently received very little educated talent, either then or later in the nineteenth century. It had to make do with the guidance handymen could give it—sometimes, of course, handymen like Henry Ford, with a dash of genius.

The curious thing was that in Germany, in the 1830's and 1840's, long before serious industrialization had started there, it was possible to get a good university education in applied science, better than anything England or the U.S. could offer for a couple of generations. I don't begin to understand this: it doesn't make social sense: but it was so. With the result that Ludwig Mond, the son of a court purveyor, went to Heidelberg and learned some sound applied chemistry. Siemens, a Prussian signals officer, at military academy and university went through what for their time were excellent courses in electrical engineering. Then they came to England, met no competition at all, brought in other educated Germans, and made fortunes exactly as though they were dealing with a rich, illiterate colonial territory. Similar fortunes were made by German technologists in the United States.

Almost everywhere, though, intellectual persons didn't comprehend what was happening. Certainly the writers didn't. Plenty of them shuddered away, as though the right course for a man of feeling was to contract out; some, like Ruskin and William Morris and Thoreau and Emerson and Lawrence, tried various kinds of fancies which were not in effect more than screams of horror. It is hard to think of a writer of high class who really stretched his imaginative sympathy, who could see at once the hideous back-streets, the smoking chimneys, the internal price—and also the prospects of life that were opening out for the poor, the intimations, up to now unknown except to the lucky, which were just coming within reach of the remaining 99 percent of his brother men. Some of the nineteenth-century Russian novelists might have done; their natures were broad enough; but they were living in a pre-industrial society and didn't have the opportunity. The only writer of world class who seems to have had an understanding of the industrial revolution was Ibsen in his old age: and there wasn't much that old man didn't understand.

For, of course, one truth is straightforward. Industrialization is the only hope of the poor. I use the word 'hope' in a crude and prosaic sense. I have not much use for the moral sensibility of anyone who is too refined to use it so. It is all very well for us, sitting pretty, to think that material standards of living don't matter all that much. It is all very well for one, as a personal choice, to reject industrialization—do a modern Walden, if you like, and if you go without much food, see most of your children die in infancy, despise the comforts of literacy, accept twenty years off your own life, then I respect you for the strength of your aesthetic revulsion. But I don't respect you in the slightest if, even passively, you try to impose the same choice on others who are not free to choose. In fact, we know what their choice would be. For, with singular unanimity, in any country where they have had the chance, the poor have walked off the land into the factories as fast as the factories could take them.

I remember talking to my grandfather when I was a child. He was a good specimen of a nineteenth-century artisan. He was highly intelligent, and he had a great deal of character. He had left school at the age of ten, and had educated himself intensely until he was an old man. He had all his class's passionate faith in education. Yet, he had never had the luck—or, as I now suspect, the worldly force and dexterity—to go very far. In fact, he never went further than maintenance foreman in a tramway depot. His life would seem to his grandchildren laborious and unrewarding almost beyond belief. But it didn't seem to him quite like that. He was much too sensible a man not to know that he hadn't been adequately used: he had too much pride not to feel a proper rancor: he was disappointed that he had not done more—and yet, compared with *his* grandfather, he felt he had done a lot. His grandfather must have been an agricultural laborer. I don't so much as know his Christian name. He was one of the 'dark people,' as the old Russian liberals used to call them, completely lost in the great anonymous sludge of history. So far as my grandfather knew, he could

not read or write. He was a man of ability, my grandfather thought; my grandfather was pretty unforgiving about what society had done, or not done, to his
ancestors, and did not romanticize their state. It was no fun being an agricultural
laborer in the mid- to late eighteenth century, in the time that we, snobs that we
are, think of only as the time of the Enlightenment and Jane Austen.

The industrial revolution looked very different according to whether one
saw it from above or below. It looks very different today according to whether
one sees it from Chelsea or from a village in Asia. To people like my grandfather,
there was no question that the industrial revolution was less bad than what had
gone before. The only question was, how to make it better.

In a more sophisticated sense, that is still the question. In the advanced
countries, we have realized in a rough and ready way what the old industrial
revolution brought with it. A great increase of population, because applied
science went hand in hand with medical science and medical care. Enough
to eat, for a similar reason. Everyone able to read and write, because an
industrial society can't work without. Health, food, education; nothing but the
industrial revolution could have spread them right down to the very poor. Those
are primary gains—there are losses too, of course, one of which is that organizing
a society for industry makes it easy to organize it for all-out war. But the gains
remain. They are the base of our social hope.

And yet: do we understand how they have happened? Have we begun to
comprehend even the old industrial revolution? Much less the new scientific
revolution in which we stand? There never was anything more necessary to
comprehend.

Recommended Readings

Anderson, A., ed., *Minds and Machines* (Englewood Cliffs, New Jersey: Prentice-Hall, 1964).

Barber, B., *Science and the Social Order* (New York: Free Press, 1952).

Bernal, J.D., *The Social Function of Science* (London: Routledge and Kegan Paul, 1939).

Cassirer, E., *An Essay on Man* (New Haven: Yale Univ. Press, 1945).

Dewey, J., *Reconstruction in Philosophy* (Boston: Beacon Press, 1948).

Kahn, H., "The Vision of Dehumanized Society," From: *The Year 2000* (New York: The Macmillan Co., 1967).

Levine, G. and Thomas, O., eds., *The Scientist vs. The Humanist* (New York: W.W. Norton, 1963).

Malinowski, B., *A Scientific Theory of Culture* (Chapel Hill, North Carolina: North Carolina Univ. Press, 1945).

Merton, R.K., *Social Theory and Social Structure* (New York: Free Press, 1957).

Polanyi, M., *Science, Faith and Society* (London: Oxford Univ. Press, 1946).

Russell, B., *The Impact of Science on Society* (New York: Simon and Schuster, 1953).

The Origins and History of Science

The Origins and History of Science

Introduction

The history of science is part of the history of man, for science is a human activity, a human institution. We chose the materials for this chapter with this in mind, endeavoring to show that the origins of science—first among the Greeks of antiquity and later in Western Europe—are complex, and are inseparable from our presuppositions about the world. In turn, our presuppositions are an integral part of the total culture, social and religious practices, economic institutions and needs, social class structures, and political organization. Hence, the question of how science came into being involves many factors which are not, strictly speaking, scientific.

The selections presented here show that the interaction of various nonscientific institutions continues to be of great importance in the rise and growth of scientific thought and practice. The church, the state, and the economy exercised enormous influence on scientific development, sometimes stimulating it, other times obstructing it. In this sense, we see the history of science as a chapter in social history. Just as social institutions influenced science, so also science, especially in modern times, influenced the course of social history. This chapter is largely concerned with this reciprocal interaction.

Several selections on the nature of scientific progress are included to establish which social and cultural conditions make science possible, or alternatively, inhibit and suppress its possibilities. Among our institutions, such as politics and government, religion and art, it appears that science alone has shown definite and significant progress. Does science "progress?" And if it does, why does it? What is the nature of its progress? Is it gradual, accumulative, and evolutionary? Or is it revolutionary and traumatic? These are among the questions discussed in the selections on scientific progress. Implicit in these considerations is the problem of the direction of science today. Are its tendencies repressive and dehumanizing? Or does it still hold out the hope of liberating mankind from unnecessary suffering and toil? This last constellation of issues is introduced in its historical perspective in this chapter, but treated in greater depth and detail in other chapters.

Essentially, we hope to show here that a knowledge of the history of science in all its rich and fascinating social and cultural relations will yield a far better grasp of the nature and importance of science today. Studies *in* science are, of course, important for the student to master if he desires a detailed knowledge of the intricacies of prevailing procedures and theories. But too much attention to the specifics of science sacrifices the opportunity to achieve a broader and more enriched overview of the field and the vast implications involved for human life and the future of all men. The present chapter is just a glimpse, or an introduction, to the many facets of the origin, nature, and growth of the scientific enterprise. It may not be totally representative and is obviously incomplete, but it is our hope that it will stimulate further reading in this interesting and extremely important field of study.

BENJAMIN FARRINGTON

Science and Politics in the Ancient World

Benjamin Farrington was Professor of Classics at University College, Swansea,
England. His many books on classical civilization and science include *Science in
Antiquity* (1936) and *Greek Science* (1944). In this selection, Farrington
describes the genesis of two remarkable achievements of Greek science before
Socrates, the atomic theory of the universe, and the beginnings of an empirical
science of medicine. These developments mark the start of a new relationship
between man and his environment characterized by a spirit of rational inquiry
and control, freed from superstitions that inhibited progress in earlier and other
contemporary civilizations.

The Atomic Theory and Hippocratic Medicine. The war on superstition begins.

The two great achievements of Ionian science before Socrates may be described
as the atomic theory of the constitution of matter, with the cosmology based on
it, and Hippocratic medicine. Leucippus and Democritus, the creators of atomism,
based their theories on a wide range of observation, but could not test the truth
of their speculations directly by experiment. Their subject-matter was inaccessible to them. The atoms, on which their whole theory rested, were by definition
too small to be the objects of sense-experience; sun, moon, and stars were
inaccessible. There was as yet no telescope, no microscope, and no science of
chemistry. Their atomism was therefore very different from the modern atomic
theory. It was a speculation based on the observation of uncontrolled natural
phenomena. The modern atomic theory, though it borrowed the concepts and
language of the old Greek speculation, differed fundamentally in being based on
data derived from controlled experiments in chemistry. It would not be correct,
however, to say that the Greeks could not appreciate the difference between
theory based on observation of natural phenomena over which they had no
control, and theory based on experiment. It was just this difference that im-
pressed itself on the Hippocratic doctors who had their material, namely, the
bodies of their patients, under their hands. The doctors were fully conscious that
every treatment they applied to a patient had its experimental as well as its
humanitarian side. And they excluded from the method of their science the
unverifiable hypotheses of the physicists.

It must not be supposed, however, that the ancient speculators on physics
were unaware of the necessity of relating every conclusion as closely as they
could to *physis*, or Nature, itself. They never lost sight of the fact that it was
Nature they were attempting to understand. Heraclitus defined wisdom as the
understanding of the way in which the universe works. The Pythagoreans were
put in the way of their special theory of the nature of things by experimenting
with the musical notes that can be drawn from taut strings, and relating the pitch

From *Science and Politics in the Ancient World* by Benjamin Farrington. 1965. Reprinted
by permission of the publishers, George Allen & Unwin Ltd., London. Pp. 57–66.

of these notes to the length of the strings. Empedocles demonstrated the corporeal nature of air by thrusting a funnel into water with the upper end closed and showing that the water could not enter until the finger was removed and the enclosed air released. When Anaxagoras wished to demonstrate that the senses have a limit beyond which their accuracy cannot be trusted, he did so by mingling a black liquid drop by drop with a white. Objectively a change in color must accompany the fall of each drop, but it is too slight to be detected by the eye. These and similar experiments show that they had taken the first step to a real technique of systematic experimental investigation, although they did not get very far with it. . . .

It can be no part of our purpose here to expound the ancient doctrine of Atomism. But it is relevant to insist that the steps by which this system was evolved, steps marked by the names of Thales, Anaximander, Anaximenes, Pythagoras, Parmenides, Zeno, Melissus, Empedocles, Anaxagoras, Leucippus and Democritus, still form an admirable introduction to scientific culture, an admirable training in rational thought. These names mark an epoch in the history of humanity. With them begins a new relation between man and his environment which, after long frustration and delay, is bearing its fruit in a fresh advance of mankind in our own day. This period gave to us for the first time in recorded history the picture of man behaving in a fully rational way in the face of nature, confident that the ways of nature were not past finding out, awed with the discovery of law in nature, freed from the superstition of animism, serene in his willing subjection to the law. The spell of this new type of man fell upon the poet Euripedes through his friendship with Anaxagoras, and he sang of it, in his choruses, to the Athenian democracy in accents that can still move us by their pregnant anticipation of what the spirit of science can mean for mankind:

Blessed is the man who has laid hold of the knowledge that comes from the enquiry into Nature. He stirreth up no evil for the citizens nor gives himself to unjust acts, but surveys the ageless order of immortal Nature, of what it is composed and how and why. In the heart of such as he the study of base acts can find no lodging.

That Euripides should have felt constrained to proclaim the political innocence of the scientist is a precious light on the temper of his time. Anaxagoras was banished from Athens for publicly teaching his scientific views.

When we turn from physics to medicine we are struck by an equal devotion to the task of observation of phenomena. . . .

. . . The Greek doctor was able to make an advance on the scientific method of the physicist. Having his material under his control, he avoided hypotheses (in the sense in which Newton used the word when he said *Hypotheses non fingo*) as much as possible, endeavoring always to submit his opinions to the test of observation. In the treatise entitled *Precepts* this point is discussed:

One must attend in medicine not primarily to plausible theories, but to experience combined with reason. For a theory is a composite memory of things apprehended by sense-perception. For the sense-perception, coming first in experience and conveying to the intellect the things subjected to it, is clearly imaged, and the intellect, receiving these things many times, noting the occasion, the time and the manner, stores them up in itself and remembers. Now I approve of theorizing, if its foundations are laid in events, and its conclusions deduced in accordance with phenomena. . . . But if it begins, not from a clear impression, but from a plausible fiction, it often induces a grievous and troublesome condition. All who follow this method are lost in a blind alley. . . . Conclusions which are merely verbal cannot bear fruit; only those do which are based on demonstrated fact. For affirmation and talk are deceptive and treacherous. Wherefore one must hold fast to facts in generalizations also, and occupy oneself with facts persistently, if one is to acquire that ready and infallible habit which we call 'the art of Medicine.'

This particular passage, being largely couched in technical Epicurean phraseology, must be dated in the third century B.C. But that it represents a tradition already two hundred years old in the old medical schools is proved by the fifth century treatise *On Ancient Medicine*. In this treatise a vigorous protest is made against the attempt to base the science of medicine on the postulates or hypotheses of Empedoclean cosmology. The physical philosophers are bidden to keep their postulates for dealing with insoluble mysteries, "for example, things in the sky or under the ground." There they are in place, "for there is no test the application of which would give confirmation." But in medicine they are not in place, "for medicine has all its material under its control." This careful discussion of the place of hypothesis in the investigation of nature is one of the landmarks in the history of ancient science, and was not without its effect on modern science, the work *On Ancient Medicine* being much studied and pondered in the sixteenth, seventeenth, and eighteenth centuries of our own era. . . .

. . . This . . . is the picture of society that seems to emerge for us as we continue the perusal of [the Hippocratic] treatise:

My own view is that those who first attributed a sacred character to this malady [epilepsy] were like the magicians, purifiers, charlatans, and quacks of our own day, men who claim great piety and superior knowledge. Being at a loss, and having no treatment which would help, they concealed and sheltered themselves behind superstition, and called this malady sacred, in order that their utter ignorance might not be exposed. . . .

But perhaps what they profess is not true, the fact being that men in need of a livelihood contrive and devise many fictions of all sorts, about this disease among other things, putting the blame for each form of the affection upon a particular god. If the patient imitate a goat, if he roar, if the convulsion is in his right side, they say that the Mother of the Gods is to blame. If he utter a loud and piercing cry, they see a resemblance to a horse and blame Poseidon!

But this disease is in my opinion no more divine than any other; it has the same nature as other diseases, and its own specific cause. . . .

This disease styled sacred comes from the same causes as others, from the things that come to and go from the body, from cold, sun, and from the changing restlessness of the winds. These things are divine. So that there is no need to put the disease in a special class and to consider it more divine than others; they are all divine and all human. Each has a nature and power of its own; none is hopeless or incapable of treatment.

The humanity of this writing is no less remarkable than its scientific spirit. This was the epoch that has bequeathed to us the composite image of the Hippocratic physician, devoted equally to the patient investigation of nature and the patient service of humanity; the healer of mind and body, with his gospel of hope that the ills of men are not supernatural punishments, but natural afflictions which knowledge in time may alleviate. They tried not to hold out false hopes. "Art is long, life is short," they repeated, enforcing the truth of Xenophanes' words, "The gods have not revealed everything to men from the beginning, but men by searching in time find out better." Meantime the search and the service were felt to be the salt of life. To one who understands, knowledge of nature and love of humanity are not two things but one. . . .

GIORGIO DE SANTILLANA

The Origins of Scientific Thought

Giorgio de Santillana is Professor of History and Philosophy of Science at Massachusetts Institute of Technology. His many books include *The Crime of Galileo* (1955) and *Reflections on Men and Ideas* (1968). De Santillana traces the decline of classical Greek science. Among the reasons for this decline, he identifies the subordination of knowledge and intellect to political ends, the rise of mysticism which met the societal needs of expanding empire and increased urbanization, economic decline and consequent widespread pessimism, and political reaction which resulted in the suppression of free inquiry.

It is a common experience of our time that enough change takes place in one generation to more than fill a century for our grandfathers. Things seem to go the other way in antiquity after about 200 B.C.—a divide which is marked not only by the death of Archimedes but by the consolidation of Roman dominion over the Hellenistic empires. Intellectually, what had been decades become centuries. Aristotle had spoken of the men who had preceded him by a hundred years as "the ancients" (*palaioi*). But writers of three and four hundred years later refer to him and his likes as we would to thinkers of a generation ago. Great changes have taken place, social, political, and philosophical. Maturity has undeniably come in, and the world is witnessing the rise of the imposing intellectual structure of Roman Law. Yet—by our standards—it is as if nothing essential had happened.

As an extreme case let us take two men who are surely independent thinkers in their own right: Simplicius and Philoponus. They not only have all the learning of the right kind, they are vigorous minds and comprehensive intellects. They think originally. Their elaborations on Plato and Aristotle have provided us with invaluable material. But they are not their successors. They live after A.D. 500, a full eight centuries later. Eight hundred years—twice the time that went from Anaximander to Archimedes. Christianity is already ruling the world—in fact, one of the two, Philoponus, is a Christian. It hardly comes to mind on reading them. Stretching a point, one might say that the next scholarly step is Eduard Zeller's great *Outlines of the History of Greek Philosophy*, published in 1860. They are not so much the last "commentators" as they are currently labeled, as early professors in the modern sense, experts in what is there no longer.

The reasons for this change of pace may concern us, too. We can discern several:

A. The Hellenistic states which came after the conquests of Alexander (and Rome is only the last of them) had not only become a very mixed civilization, they had become uniformly "big-time," with huge and fearsome structures of power. State cults of divine rulers, state-encouraged superstitions, the worship of blind Fortune, had replaced the old city gods. In their wake had come new mystery cults from the East, exotic gospels of salvation, magic doctrines and practices and novelties for big-city dwellers—very much as these things have come in our time to southern California. Science was represented no longer by free men and respected elders of the community, but by subsidized intellectuals who were told to go and perform and quote one another in ample institutions like the library of Alexandria, and also (this was an order) to provide moral uplift and entertainment for the arts-loving ruling class. The insistence, which to us seems excessive, on virtue and character-building is not all escape literature, witness Tacitus and Marcus Aurelius. But there is a ring of frustration in it.

B. Economic decline has set in. There seems to be a failure of imagination at the root of it all. A great and stable and ever more complicated administration needs economic growth to keep pace with it, and the Roman Empire seems to have been strangely incapable of economic and political growth; even more so than the Chinese. Urban luxury, bread-and-games, world-wide services, the unending attrition of border wars, the subsidizing of intractable tribes, are forms of conspicuous waste which cannot rest on merely agrarian economies and growing masses of plantation slave labor. The Romans never developed, as the Middle Ages did, a system of credit banking. The medium remained coin, and the deficit and the needs for an increased circulation were met by usury and debasing of the coinage. The burden of taxes became so crushing that farmers fled from their land, had to be brought back in chains as "tied to the sod." Growing regulations meant increasing avoidance, growing expenditure, harsher enforcement, in a never-ending circle. Workers trying to flee over the border were sentenced to death by fire. Dullness, conformity, and gloom spread like a pall of smog over the last centuries. Science became manuals and encyclopedias, literature became stale rhetoric on "classic" models, or tales of the wondrous.

C. The failure of imagination explains, among other things, why men became so reactionary-minded, even when they thought they were entertaining the most lofty and liberal ideals. Something like that was to occur again in the American South. When Aristotle, the great master of ethics, said that slavery is a fact of nature, and that we shall need slaves so long as the shuttle will not run in the loom by itself, he had registered one of those great mental blocks which foretell the end of a cycle. And this leads us to what is obviously crucial, the lack of an applied science.

Pure science is always a hazardous and unfinished affair, stretching out its structures in perilous balance over the unknown. It does not suit men's whims or comfort their fears. In order to be accepted by a tough-minded society, it must produce unquestionable and stunning results, as happened with Newton's laws. Otherwise, it will be told to lay off and not disturb people's minds unnecessarily. Men like Galileo, when they dare to speak openly, will be reproved. It happened at the freest moment of Greek thought with Anaxagoras; it happened again in a different context with Aristarchus and his Copernican suggestion. Much has been said of a "loss of nerve" in Greek speculation after 300 B.C. The expression may not be accurate, but it circumscribes something that certainly took place: an inflection away from certain lines of research, a lack of aggressiveness, a kind of settling down.

The decisive point lies in the use of mathematics. After the early efforts of the Pythagoreans to set up a mathematical physics, it is as if the enterprise were abandoned as unfruitful. The turning point lies within the career of Eudoxus, who was Archytas's pupil. It is, as we said earlier, as if the lifting power of pure number had been finally released to rise into the realm of abstraction, providing henceforth nothing but transposed mathematical models for reality. It "saved" the phenomena instead of explaining them. The spheres of Eudoxus, Plato's *Timaeus* as a whole, are such models. The relation of number to reality has been transformed irretrievably, for even Archimedes is unable to reverse the trend. The higher qualitative virtues of number, so to speak, have been saved at the expense of actual application.

Attention was drawn some years ago by Werner Jaeger to the fact that, beginning with Isocrates the orator, one of Aristotle's teachers, there are thinkers deliberately opposed to *akribeia*, precision in measurement and precision in general. Aristotle's philosophy is in itself, as we have seen, a turning away from mathematics to other metaphysically more "relevant" forms of accuracy. This did not preclude measurement as such, witness Dicaiarchos's (Aristotle's own pupil's) enterprise in measuring the height of mountains, but demoted it, in a way, to philosophical irrelevance. The issue verged substantially on the existence of physical truths not susceptible of measurement, and hence, whenever other reasons so suggested, on the possibility of ignoring quantitative data. A disturbing example comes to us from mathematical geography. The name of Ptolemy is synonymous for us with the greatest effort of antiquity at astronomical exactitude. Ptolemy's respect for precision in heavenly movements even led him into painful and unesthetic complications, which Copernicus was later to hold against him. Yet the same Ptolemy, when it came to establishing a mathematical geography on the available data, was willing to "adjust," if not the data, at least the inferences. The size of the Western Ocean appeared on the spherical map as disproportionately large, extending as it did unbrokenly from the Canaries to China. Ptolemy tried to compromise by extending all the dimensions of land (part dimly known) from east to west and taking the Stoic position that China must be just half the globe away from the Canaries.

The publication of Ptolemy's *Geography* in the Renaissance was to cause no small surprise among Italian scholars, and the problem of the size of the earth and of the exact value of ancient measures became an issue; the debates were a stimulus to Columbus's enterprise. Thus we find, even in this most practical of fields, an overriding concern, that we may call theoretical only if we give the word *theōria* its original meaning.

The relationship of Greek thought to nature remains fundamentally different from ours: it is not a search for the point of attack from which to attempt a break-through, but a quest for harmony, proportion, an over-all order to which to adjust. Man sees himself as living *with* nature, not opposite to nature, a member of the great republic of gods, men, and all that is. From the very start we feel the persistent effort to justify the cosmos, to rid its powers of their dark and dreadful ambiguity. Order, fitness, justice, reconciliation, reassurance, are the themes which occur at times in the moral key, but always showing reason, the clarity of truth, *alētheia*, as the saving power against terrified animal subjection. In Democritus, the resolute physicist, we find again "a mind devoid of fear" as the highest good. The very word for science is *epistēmē,* derived from a verb which means "standing up to." This meaning may be more helpful than many economic theories to explain why science in Greece was not as subject to the pressure for practical results as it was from the Renaissance to our days. In the Anaximandrian equation the central symbol is "fittingness"; in the Pythagorean *mathēsis* or "doctrine," the spiritual tensions are reconciled in a scheme of intellectual salvation. We base our thinking on nature as necessity. For the Greek, necessity is not quite of nature, it is seen as part of the purifying vision: the logic of a theorem, the rigorous timing of the celestial orbs, will be felt as analogous to the determination of holy ritual. The Logos determines, rules, makes clear; it does not by itself necessitate. Galileo will insist that the universe is ruled by the necessity of mathematical law, but the Pythagoreans would have mentioned Reason alone, and even Democritus, we remember, said it is ruled by Reason *and* Necessity, as if they were two different entities. Necessity and Reason are closely coupled since Anaximander, but they remain formally distinct even with the Stoics, complementary as it were. The dark mythical figure of Necessity, *Anankē*, is imagined as a kind of inchoate heavy passiveness, which precedes the order of *physis* and reason; when Thucydides uses the word he means the massive inertia whereby events keep going in a certain direction once they have been set in motion. Behaviors and determinations come from the inherent reason: "Nature," says Aristotle in a revealing phrase, "refuses to be badly administered." Wonderfully said, but it implies that it is administered from a higher order, and then we cannot expect the objects of nature to behave according to inherent laws; we can hardly think of questioning them by way of experiment.

The later subjection of the astronomical profession to the dictates of the philosophers is, so to speak, inscribed in the very origins; it expresses a hierarchical concept which exists orginally inside the single creative mind. In Hellenistic times it became an organizational split which was to be healed only by Galileo and Kepler, who created the figure of the "astronomer phylosophicall." It is some such reason, apparently, which precluded the application of mathematics to terrestrial motion and change. Archimedes' statics, based on symmetry and proportion, remains inside the classic frame; it does not point the way to the entirely new complex of ideas still to be born. Thus the ancient world remains without a science of dynamics, and that may be not the last cause of its downfall. An accepted explanation for the failure to use steam power in antiquity (Heron's steam jet reactor remained a toy) is the lack of a craftsman class specialized in precision machining of metal. After the reconstruction of the Antikythera machine, this explanation does not look persuasive. The men who built that machine were fully the equals of Harrison and Watt. They could have built a chronometer: it is symbolic that their skill was used on computing astronomical cycles and epicycles. But there is worse. Most of the inventions displayed by Heron in applied mechanics are aimed only at catching the eye of the public. They are automata and trick mechanisms meant for parks, palaces, and temples, very much like those which were developed in our eighteenth century for the amusement of a mercantile aristocracy. What was really lacking were the inceptors of the scientific revolution, idea-men like Galileo, Huygens, Newton and Bernoulli. For the lack of such, ancient capitalism remained mercantile and usurary, and ultimately foundered on its own slave labor.

The achievements of Greek science, such as they were, did provide the starting point for our scientific rebirth in the seventeenth century. But the spirit of those achievements remained embodied, during the intervening period, in a few great scientific-philosophical doctrines which could claim a true religious content, and thus not only guided but gave shape to Western thought in the difficult time of transition. They were, one might say, the carrier waves. It is through them that the continuity was preserved, and without them further developments would remain incomprehensible. We shall conclude, therefore, with a sketch of three forms of scientific religion: Atomism, Stoicism, and Neoplatonism.

DIRK STRUIK

A Concise History of Mathematics

Dirk Struik was Professor of Mathematics at Massachusetts Institute of Technology. He published many books and articles on mathematics and its history, including his widely studied *Concise History of Mathematics* (1948). In the selections concerning the rise and decline of Greek science we find that social, political, and economic factors were often decisively important in the progress and development of scientific thought. In the essay you are about to read, Struik shows that these "external" factors were of great significance in the origin and dissemination of modern mathematical concepts and procedures. The demise of the Roman empire and the advent of feudalism inhibited the growth of mathematical knowledge after it had reached a point of relative sophistication among the ancients. This essay, among others in this volume, emphasizes the importance of studying the social relations of science as a way of better understanding its function in our own culture.

The most advanced section of the Roman Empire from both an economic and a cultural point of view had always been the East. The Western part had never been based on an irrigation economy; its agriculture was of the extensive kind which did not stimulate the study of astronomy. Actually the West managed very well in its own way with a minimum of astronomy, some practical arithmetic, and some mensuration for commerce and surveying; but the stimulus to promote these sciences came from the East. When East and West separated politically this stimulation almost disappeared. The static civilization of the Western Roman Empire continued with little interruption or variation for many centuries; the Mediterranean unity of antique civilization also remained unchanged—and was not even very much affected by the barbaric conquests. In all Germanic kingdoms, except perhaps those of Britain, the economic conditions, the social institutions, and the intellectual life remained fundamentally what they had been in the declining Roman Empire. The basis of economic life was agriculture, with slaves gradually replaced by free and tenant farmers; but in addition there were prosperous cities and a large-scale commerce with a money economy. The central authority in the Greco-Roman world after the fall of the Western Empire in 476 was shared by the emperor in Constantinople and the popes of Rome. The Catholic Church of the West through its institutions and language continued as best it could the cultural tradition of the Roman Empire among Germanic kingdoms. Monasteries and cultured laymen kept some of the Greco-Roman civilization alive.

One of these laymen, the diplomat and philosopher Anicius Manlius Severinus Boetius, wrote mathematical texts which were considered authoritative in the Western world for more than a thousand years. They reflect the cultural conditions, for they are poor in content and their very survival may have been

From *A Concise History of Mathematics* by Dirk Struik. Dover Publications, Inc., New York, 1967. Reprinted through the permission of the publisher. Pp. 83–93.

influenced by the belief that the author died in 524 as a martyr to the Catholic
faith. His *Institutio arithmetica,* a superficial translation of Nicomachus, did
provide some Pythagorean number theory which was absorbed in medieval
instruction as part of the ancient trivium and quadrivium: arithmetic, geometry,
astronomy, and music.

It is difficult to establish the period in the West in which the economy of the
ancient Roman Empire disappeared to make room for a new feudal order. Some
light on this question may be shed by the hypothesis of H. Pirenne,[1] according
to which the end of the ancient Western world came with the expansion of Islam.
The Arabs dispossessed the Byzantine Empire of all its provinces on the Eastern
and Southern shores of the Mediterranean and made the Eastern Mediterranean a
closed Muslim lake. They made commercial relations between the Near Orient
and the Christian Occident extremely difficult for several centuries. The intellectual avenue between the Arabic world and the Northern parts of the former
Roman Empire, though never wholly closed, was obstructed for centuries.

Then in Frankish Gaul and other former parts of the Roman Empire large-
scale economy subsequently vanished; decadence overtook the cities; returns
from tolls became insignificant. Money economy was replaced by barter and local
marketing. Western Europe, in short, was reduced to a state of semi-barbarism.
The landed aristocracy rose in significance with the decline of commerce; the
North Frankish landlords, headed by the Carolingians, became the ruling power
in the land of the Franks. The economic and cultural center moved to Northern
France and Britain. The separation of East and West limited the effective
authority of the Pope to the extent that the papacy allied itself with the Carolingians, a move symbolized by the crowning of Charlemagne in 800 as Emperor of
the Holy Roman Empire. Western society became feudal and ecclesiastical, its
orientation Northern and Germanic.

During the early centuries of Western feudalism we find little appreciation of
mathematics even in the monasteries. In the again primitive agricultural society of
this period the factors stimulating mathematics, even of a directly practical kind,
were nearly nonexistent; and monasteric mathematics was no more than some
ecclesiastical arithmetic used mainly for the computation of Easter-time (the so-
called *computus*). Boetius was the highest source of authority. Of some importance among these ecclesiastical mathematicians was the British-born Alcuin,
associated with the court of Charlemagne, whose *Problems for the Quickening of
the Mind of the Young* contained a selection which have influenced the writers
of textbooks for many centuries. Many of these problems date back to the
ancient Orient. For example:

A dog chasing a rabbit, which has a start of 150 feet, jumps 9 feet every time
the rabbit jumps 7. In how many leaps does the dog overtake the rabbit?
A wolf, a goat, and a cabbage must be moved across a river in a boat holding
only one beside the ferry man. How must he carry them across so that the goat
shall not eat the cabbage, nor the wolf the goat?

Another ecclesiastical mathematician was Gerbert, a French monk, who in 999 became Pope under the name of Sylvester II. He wrote some treatises under the influence of Boetius, but his chief importance as a mathematician lies in the fact that he was one of the first Western scholars who went to Spain and made studies of the mathematics of the Arabic world.

There are significant differences between the development of Western, early Greek, and Oriental feudalism. The extensive character of Western agriculture made a vast system of bureaucratic administrators superfluous, so that it could not supply a basis for an eventual Oriental despotism. There was no possibility in the West of obtaining vast supplies of slaves. When villages in Western Europe grew into towns, these towns developed into self-governing units, in which the burghers were unable to establish a life of leisure based on slavery. This is one of the main reasons why the development of the Greek *polis* and the Western city, which during the early stages had much in common, deviated sharply in later periods. The medieval townspeople had to rely on their own inventive genius to improve their standard of living. Fighting a bitter struggle against the feudal landlords—and with much civil strife in addition—they emerged victorious in the twelfth, thirteenth, and fourteenth centuries. This triumph was based not only on a rapid expansion of trade and money economy, but also on a gradual improvement in technology. The feudal princes often supported the cities in their fight against the smaller landlords, and then eventually extended their rule over the cities. This finally led to the emergence of the first national states in Western Europe.

The cities began to establish commercial relations with the Orient, which was still the center of civilization. Sometimes these relations were established in a peaceful way, sometimes by violent means as in the many Crusades. First to establish mercantile relations were the Italian cities; they were followed by those of France and Central Europe. Scholars followed, or sometimes preceded, the merchant and the soldier. Spain and Sicily were the nearest points of contact between East and West, and here Western merchants and students became acquainted with Islamic civilization. When in 1085 Toledo was taken from the Moors by the Christians, Western students flocked to this city to learn science as it was transmitted in Arabic. They often employed Jewish interpreters to converse and to translate, and so we find in twelfth-century Spain, Plato of Tivoli, Gherardo of Cremona, Adelard of Bath, and Robert of Chester, translating Arabic mathematical manuscripts into Latin. Thus Europe became familiar with Greek classics through the Arabic; and by this time Western Europe was advanced enough to appreciate this knowledge.

As we have said, the first powerful commercial cities arose in Italy, where during the twelfth and thirteenth centuries Genoa, Pisa, Venice, Milan, and Florence carried on a flourishing trade between the Arabic world and the North. Italian merchants visited the Orient and studied its civilization; Marco Polo's travels show the intrepidity of these adventurers. Like the Ionian merchants of almost two thousand years before, they tried to study the science and the arts of the older civilization not only to reproduce them, but also to assimilate them into their own mercantile society, which already in the twelfth and thirteenth centuries saw the growth of banking and the beginnings of a capitalist form of industry. The first Western merchant whose mathematical studies showed a certain maturity was Leonardo of Pisa.

Leonardo, also called Fibonacci ("son of Bonaccio"), traveled in the Orient as a merchant. On his return he wrote his *Liber Abaci* (1202), filled with arithmetical and algebraical information which he had collected on his travels. In the *Practica Geometriae* (1220), Leonardo described in a similar way whatever he had discovered in geometry and trigonometry. He may have been an original investigator as well, since his books contain many examples which seem to have no exact duplicates in Arabic literature.[2] . . .

The *Liber Abaci* is one of the means by which the Hindu-Arabic system of numeration was introduced into Western Europe. Its occasional use dates back to centuries before Leonardo, when it was imported by merchants, ambassadors, scholars, pilgrims, and soldiers coming from Spain and from the Levant. The oldest dated European manuscript containing the numerals is the *Codex Vigilanus,* written in Spain in 976. However, the introduction of the ten symbols into Western Europe was slow; the earliest French manuscript in which they are found dates from 1275. The Greek system of numeration remained in vogue along the Adriatic for many centuries. Computation was often performed on the ancient abacus, a board with counters or pebbles (often simply consisting of lines drawn in sand) similar in principle to the counting boards still used by the Russians, Chinese, Japanese, and by children on their baby-pens. Roman numerals were used to registrate the result of a computation on the abacus. Throughout the Middle Ages (and even later) we find Roman numerals in merchant's ledgers, which indicates that the abacus was used in the offices. The introduction of Hindu-Arabic numerals met with opposition from the public, since the use of these symbols made merchant's books difficult to read. In the statutes of the *Arte del Cambio* of 1299 the bankers of Florence were forbidden to use Arabic numerals and were obliged to use cursive Roman ones. Sometime during the fourteenth century Italian merchants began to use some Arabic figures in their account books.[3] Occasionally we find intermediate forms such as $II^m III^c XV$ for 2315.

With the extension of trade, interest in mathematics spread slowly to the Northern cities. It was at first mainly a practical interest, and for several centuries arithmetic and algebra were taught outside the universities by self made reckon masters, usually ignorant of the classics, who also taught bookkeeping and navigation. For a long time this type of mathematics kept definite traces of its Arabic origin, as words such as "algebra" and "algorithm" testify.

Speculative mathematics did not entirely die during the Middle Ages, though it was cultivated not among the men of practice, but among the scholastic philosophers. Here the study of Plato and Aristotle, combined with meditations on the nature of the Deity, led to subtle speculations on the nature of motion, of the continuum and of infinity. Origen had followed Aristotle in denying the existence of the actually infinite, but St. Augustine in the *Civitas Dei* had accepted the whole sequence of integers as an actual infinity. His words were so well chosen that Georg Cantor has remarked that the transfinitum cannot be more energetically desired and cannot be more perfectly determined and defended than was done by St. Augustine.[4] The scholastic writers of the Middle Ages, especially St. Thomas Aquinas, accepted Aristotle's *infinitum actu non datur,*[5] but considered every continuum as potentially divisible ad infinitum. Thus there was no smallest line. A point, therefore, was not a part of a line, because it was indivisible: *ex indivisilibus non potest compari aliquod continuum.*[6] A point could generate a line by motion. Such speculations had their influence on the inventors of the infinitesimal calculus in the seventeenth century and on the philosophers of the transfinite in the nineteenth; Cavalieri, Tacquet, Bolzano, and Cantor knew the scholastic authors and pondered over the meaning of their ideas. . . .

The main line of mathematical advance passed through the growing mercantile cities under the direct influence of trade, navigation, astronomy, and surveying. The townspeople were interested in counting, in arithmetic, in computation. Sombart had labeled this interest of the fifteenth- and sixteenth-century burgher his *Rechenhaftigkeit.*[7] Leaders in the love for practical mathematics were the reckon masters, only very occasionally joined by a university man, able, through his study of astronomy, to understand the importance of improving computational methods. Centers of the new life were the Italian cities and the Central European cities of Nuremberg, Vienna, and Prague. The fall of Constantinople in 1453, which ended the Byzantine Empire, led many Greek scholars to the Western cities. Interest in the original Greek texts increased, and it became easier to satisfy this interest. University professors joined with cultured laymen in studying the texts, ambitious reckon masters listened and tried to understand the new knowledge in their own way.

   Typical of this period was Johannes Müller of Königsberg, or Regiomontanus, the leading mathematical figure of the fifteenth century. The activity of this remarkable computer, instrument maker, printer, and scientist illustrates the advances made in European mathematics during the two centuries after Leonardo. He was active in translating and publishing the classical mathematical manuscripts available. His teacher, the Viennese astronomer, George Peurbach—author of astronomical and trigonometrical tables—had already begun a translation of the astronomy of Ptolemy from the Greek. Regiomontanus continued this translation and also translated Appolonius, Heron, and the most difficult of all, Archimedes. His main original work was *De triangulis omnimodus libri quinque* (1464, not printed until 1533), a complete introduction into trigonometry, differing from our present-day texts primarily in the fact that our convenient notation did not exist. It contains the law of sines in a spherical triangle. All theorems had still to be expressed in words. Trigonometry, from that point on, became a science independent of astronomy. Nasīr al-din had accomplished something similar in the thirteenth century, but it is significant that his work never resulted in much further progress, whereas Regiomontanus' book deeply influenced further development of trigonometry and its application to astronomy and algebra. Regiomontanus also devoted much effort to the computation of trigonometric tables. He has, for instance, tables of sines to radius 60.000 for intervals of one minute which were printed after his death.

Sines were line segments, defined as semichords subtending angles in a circle. Their numerical values therefore depended on the length of the radius. A large radius allowed great accuracy in the value of the sines, without the necessity of introducing sexagesimal (or decimal) fractions. The systematic use of radius 1, and hence the concept of sines, tangents, etc., as ratios (numbers) is due to Euler (1748).

So far no definite step had been taken beyond the ancient achievements of the Greeks and Arabs. The classics remained the *ne plus ultra* of science. It came therefore as an enormous and exhilarating surprise when Italian mathematicians of the early sixteenth century actually showed that it was possible to develop a new mathematical theory which the Ancients and Arabs had missed. This theory, which led to the general algebraic solution of the cubic equation, was discovered by Scipio del Ferro and his pupils at the University of Bologna.

The Italian cities had continued to show proficiency in mathematics after the time of Leonardo. In the fifteenth century their reckon masters were well versed in arithmetical operations, including surds (without having any geometrical scruples), and their painters were good geometers. Vasari, in his *Lives of the Painters,* stresses the considerable interest which many quattrocento artists showed in solid geometry. One of their achievements was the development of perspective by such men as Alberti and Piero della Francesca; the latter also wrote a volume on regular solids. The reckon masters found their interpreter in the Franciscan monk Luca Pacioli, whose *Summa de Arithmetica* was printed in 1494—one of the first mathematical books to be printed.[8] Written in Italian— and not a very pleasant Italian—it contained all that was known in that day of arithmetic, algebra, and trigonometry. By then the use of Hindu-Arabic numerals was well established, and the arithmetical notation did not greatly differ from ours. . . .

At this point began the work of the mathematicians at the University of Bologna. This university, around the turn of the fifteenth century, was one of the largest and most famous in Europe. Its faculty of astronomy alone at one time had sixteen lectors. From all parts of Europe students flocked to listen to the lectures—and to the public disputations which also attracted the attention of large, sportively-minded crowds. Among the students at one time or another were Pacioli, Albrecht Dürer, and Copernicus. Characteristic of the new age was the desire not only to absorb classical information but also to create new things, to penetrate beyond the boundaries set by the classics. The art of printing and the discovery of America were examples of such possibilities. Was it possible to create new mathematics? Greeks and Orientals had tried their ingenuity on the solution of the third degree equation but had only solved some special cases numerically. The Bolognese mathematicians now tried to find the general solution. . . .

Algebra and computational arithmetic remained for many decades the favorite subject of mathematical experimentation. Stimulation no longer came only from the *Rechenhaftigkeit* of the mercantile bourgeoisie but also from the demands made on surveying and navigation by the leaders of the new national states. Engineers were needed for the erection of public works and for military constructions. Astronomy remained, as in all previous periods, an important domain for mathematical studies. It was the period of the great astronomical theories of Copernicus, Tycho Brahe, and Kepler. A new conception of the universe emerged.

Philosophical thought reflected the trends in scientific thinking; Plato with his admiration for quantitative mathematical reasoning gained ascendancy over Aristotle. Platonic influence is particularly evident in Kepler's work. Trigonometrical and astronomical tables appeared with increasing accuracy, especially in Germany. The tables of G.J. Rheticus, finished in 1596 by his pupil Valentin Otho, contain the values of all six trigonometric values for every ten seconds to ten places. The tables of Pitiscus (1613) went up to fifteen places. The technique of solving equations and the understanding of the nature of their roots also improved. The public challenge, made in 1593 by the Belgian mathematician Adriaen van Roomen, to solve the equation of the 45th degree . . . was characteristic of the times. . . .

Bibliography

1. H. Pirenne, *Mahomet et Charlemagne* (Paris, 1937; English translation, New York, 1939). See A.F. Havighurst, *The Pirenne Thesis* (Boston: D.C. Heath & Company, 1958).
2. L.C. Karpinski [*Amer. Math. Monthly,* Vol. 21 (1914), pp. 37–48], using the Paris manuscript of Abū Kāmil's algebra, claims that Leonardo followed Abū Kāmil in a whole series of problems.
3. In the Medici account books (dating from 1406) of the Selfridge Collection on deposit at the Harvard Graduate School of Business Administration, Hindu-Arabic numerals frequently appear in the narrative or descriptive column. From 1439 onward they replace Roman numerals in the money or effective column of the books of primary entry—journals, wastebooks, etc.–but not until 1482 were Roman numerals abandoned in the money column of the business ledgers of all but one Medici merchant. From 1494, only Hindu-Arabic numerals are used in all the Medici account books. (From a letter by Dr. Florence Edler De Roover.) See also, F. Edler, *Glossary of Medieval Terms of Business* (Cambridge, Mass., 1934), p. 389.
4. G. Cantor, "Letter to Eulenburg (1886)," *Ges. Abhandlungen* (Berlin, 1932), pp. 401–02. The passage quoted by Cantor, Ch. XVIII of Book XII of *The City of God* (in the Healey translation) is entitled "Against such as say that things infinite are above God's knowledge."
5. "There is no actually infinite." See further E. Bodewig, "Die Stellung des hl. Thomas von Aquino zur Mathematik," *Archiv. f. Geschichte der Philosophie*, Vol 41 (1932), pp. 408–34.
6. "A continuum cannot consist of indivisibles."
7. W. Sombart, *Der Bourgeois* (Munich, Leipzig, 1913), p. 164. The term *Rechenhaftigkeit* indicates a willingness to compute, a belief in the usefulness of arithmetical work.
8. The first printed mathematical books were a commercial arithmetic (Treviso, 1478) and a Latin edition of Euclid's *Elements* (Venice, Ratdolt, 1482).

EDGAR ZILSEL

The Sociological Roots of Science

Edgar Zilsel came to the United States from Germany in the 1940's. He then
published a number of brilliant articles on the history and philosophy of science
in leading scholarly journals. The selection below is taken from this period.
Zilsel's contribution to this volume sets the stage for the rise of science as we
know it today. Here is a concise account of the social foundations of, perhaps,
the most powerful and far-reaching institution in human history: the successful
rise and spread of modern European science. Why modern science appeared
at this particular time in history is the subject of the essay. Zilsel's thesis is
that science was born during the progress of technology, when enthusiasm
for mechanical arts and experimental methods overcame the social prejudice
against manual labor and was adopted by rationally trained scholars around
the year 1600.

It is not impossible . . . to study the emergence of modern science as a sociological process. Since this emergence took place in the period of early European
capitalism, we shall have to review that period from the end of the Middle Ages
until 1600. Certain stages of the scientific spirit, however, developed in other
cultures too, e.g., in classical antiquity and, to a lesser degree, in some Oriental
civilizations and in the Arabic culture of the Middle Ages. Moreover, the scientific
cultures are not independent of each other. In modern Europe the beginnings of
science, particularly, have been greatly influenced by the achievements of ancient
mathematicians and astronomers and medieval Arabic physicians. We shall, however, discuss not this influence but the sociological conditions which made it
possible. We can, necessarily, give but a sketchy and greatly simplified analysis of
this topic here. All details and much of the evidence must be left to a more
extensive exposition at another place.

Human society has not often changed so fundamentally as it did with the transition from feudalism to early capitalism. These changes are generally known. Even
in a very brief exposition of the problem, however, we must mention some of
them, since they form necessary conditions for the rise of science.

   1. The emergence of early capitalism is connected with a change in both the
setting and the bearers of culture. In the feudal society of the Middle Ages the
castles of knights and rural monasteries were the centers of culture. In early
capitalism culture was centered in towns. The spirit of science is worldly and not
military. Obviously, therefore, it could not develop among clergymen and knights
but only among townspeople.

   2. The end of the Middle Ages was a period of rapidly progressing technology
and technological inventions. Machines began to be used both in production of
goods and in warfare. On the one hand, this set tasks for mechanics and chemistry,

Edgar Zilsel, "The Sociological Roots of Science." *American Journal of Sociology,* January
1942, pp. 545–560. Copyright 1942 by The University of Chicago.

and, on the other, it furthered causal thinking, and, in general, weakened magical thinking.

3. In medieval society the individual was bound to the traditions of the group to which he unalterably belonged. In early capitalism economic success depended on the spirit of enterprise of the individual. In early feudalism economic competion was unknown. When it started among the craftsmen and tradesmen of the late medieval towns, their guilds tried to check it. But competition proved stronger than the guilds. It dissolved the organizations and destroyed the collective-mindedness of the Middle Ages. The merchant or craftsman of early capitalism who worked in the same way as his fathers had was outstripped by less conservative competitors. The individualism of the new society is a presupposition of scientific thinking. The scientist, too, relies, in the last resort, only on his own eyes and his own brain and is supposed to make himself independent of belief in authorities. Without criticism there is no science. The critical scientific spirit (which is entirely unknown to all societies without economic competition) is the most powerful explosive human society ever has produced. If the critical spirit expanded to the whole field of thinking and acting it would lead to anarchism and social disintegration. In ordinary life this is prevented by social instincts and social necessities. In science itself the individualistic tendencies are counterbalanced by scientific cooperation. This, however, will be discussed later.

4. Feudal society was ruled by tradition and custom, whereas early capitalism proceeded rationally. It calculated and measured, introduced bookkeeping, and used machines. The rise of economic rationality furthered development of rational scientific methods. The emergence of the quantitative method, which is virtually nonexistent in medieval theories, cannot be separated from the counting and calculating spirit of capitalistic economy. The first literary exposition of the technique of double-entry bookkeeping is contained in the best textbook on mathematics of the fifteenth century, Luca Pacioli's *Summa de arithmetica* (Venice, 1494); the first application of double-entry bookkeeping to the problems of public finances and administration was made in the collected mathematical works of Simon Stevin, the pioneer of scientific mechanics (*Hypomnemata mathematica* [Leyden, 1608]), and a paper of Copernicus on monetary reform (*Monetae cudendae ratio* [composed in 1552]) is among the earliest investigations of coinage. This cannot be mere coincidence.

The development of the most rational of sciences, mathematics, is particularly closely linked with the advance of rationality in technology and economy. The modern sign of mathematical equality was first used in an arithmetical textbook of Recorde that is dedicated to the "governors and the reste of the Companio of Venturers into Moscovia" with the wish for "continualle increase of commoditie by their travell" (*The Wetstone of Witte* [London, 1557]). Decimal fractions were first introduced in a mathematical pamphlet of Stevin that begins with the words: "To all astronomers, surveyors, measurers of tapestry, barrels and other things, to all mintmasters and merchants good luck!" (*De thiende* [Leyden, 1585]). Apart from infusions of Pythagorean and Platonic metaphysics, the mathematical writings of the fifteenth and sixteenth centuries first deal in detail with problems of commercial arithmetic and, second, with the technological needs of military engineers, surveyors, architects, and artisans. The geometrical and arithmetical treatises of Piero de' Franceschi, Luca Pacioli, and Tartaglia in Italy, Recorde and Leonard Digges in England, Dürer and Stifel in Germany, are cases in point. Classical mathematical tradition (Euclid, Archimedes, Apollonius, Diophantus) could be revived in the sixteenth century because the new society had grown to demand calculation and measurement.

Even rationalization of public administration and law had its counterpart in scientific ideas. The loose state of feudalism with its vague traditional law was gradually superseded by absolute monarchies with central sovereignty and rational statute law. This political and juridical change promoted the emergence of the idea that all physical processes are governed by rational natural laws established by God. This, however, did not occur before the seventeenth century (Descartes, Huyghens, Boyle).

We have mentioned a few general characteristics of early capitalistic society which form necessary conditions for the rise of the scientific spirit. In order to understand this development sociologically, we have to distinguish three strata of intellectual activity in the period from 1300 to 1600: the universities, humanism, and labor.

At the universities theology and scholasticism still predominated. The university scholars were trained to think rationally but exercised the methods of scholastic rationalism which differ basically from the rational methods of a developed economy. Tradesmen are interested in reckoning; craftsmen and engineers in rational rules of operation, in rational investigation of causes, in rational physical laws. Schoolteachers, on the other hand, take an interest in rational distinction and classifications. The old sentence, *bene docet qui bene distinguit,* is as correct as it is sociologically significant. Schoolteaching, by its sociological conditions, produces a specific kind of rationality, which appears in similar forms wherever old priests, intrusted with the task of instructing priest candidates, rationalize vague and contradictory mythological traditions of the past. Brahmans in India, Buddhist theologians in Japan, Arabic and Catholic medieval scholastics conform in their methods to an astonishing degree. Jewish Talmudists proceeded in the same way, though, not being priests by profession, they dealt with ritual and canon law rather than with proper theological questions. This school rationality has developed to a monstrous degree in Brahmanic Sankhya-philosophy (sankhya means "enumeration").

As a rule the specific scholastic methods are preserved when theologians, in the course of social development, apply themselves to secular subject matters. Thus in Indian literature Brahmans who had entered the service of princes discussed politics and erotics by meticulously distinguishing and enumerating the various possibilities of political and sexual life (Kautilya, Vatsyayana). In a somewhat analogous way the medieval scholastics and the European university scholars before 1600 indulged in subtle distinctions, enumerations, and disputations. Bound to authorities, they favored quotations and uttered their opinions for the most part in the form of commentaries and compilations. After the thirteenth century mundane subject matters were treated by scholars, too, and, as an exception, even experience was referred to by some of them. But when the Schoolmen were at all concerned with secular events they did not, as a rule, investigate causes and, never, physical laws. They endeavored rather to explain the ends and meanings of the phenomena. Obviously, the occult qualities and Aristotelian substantial forms of scholasticism are but rationalizations of pre-scientific, magic, and animistic teleology. Thus till the middle of the sixteenth century the universities were scarcely influenced by the development of contemporary technology and by humanism. Their spirit was still substantially medieval. It seems to be a general sociological phenomenon that rigidly organized schools are able to offer considerable resistance to social changes of the external world.

The first representatives of secular learning appeared in the fourteenth century in Italian cities. They were not scientists but secretaries and officials of municipalities, princes, and the pope looking up with envy to the political and cultural achievements of the classical past. These learned officials who chiefly had to conduct the foreign affairs of their employers became the fathers of humanism. Their aims derive from the conditions of their profession. The more erudite and polished their writings, the more eloquent their speeches, the more prestige redounded to their employers and the more fame to themselves. They therefore chiefly strove after perfection of style and accumulation of classical knowledge. In the following centuries the Italian humanists lost in large part their official connections. Many became free literati, dependent on princes, noblemen, and bankers as patrons. Others were engaged as instructors to the sons of princes, and several got academic chairs and taught Latin and Greek at universities. Their aims remained unchanged, and their pride of memory and learning, their passion for fame, even increased. They acknowledged certain ancient writers as patterns of style and were bound to these secular authorities almost as strictly as the theologians were to their religious ones. Though humanism also proceeded rationally, its methods were as different from scholastic as from modern scientific rationality. Humanism developed the methods of scientific philology, but neglected causal research and was ignorant of physical laws and quantitative investigation. Altogether it was considerably more interested in words than in things, more in literary forms than in contents. Humanism spread over all parts of western and central Europe. Though the professional conditions and intellectual aims of the humanists outside Italy were somewhat more complex, on the whole their methods were the same.

The university scholars and the humanistic literati of the Renaissance were exceedingly proud of their social rank. Both disdained uneducated people. They avoided the vernacular and wrote and spoke Latin only. Further, they were attached to the upper classes, sharing the social prejudices of the nobility and the rich merchants and bankers and despising manual labor. Both, therefore, adopted the ancient distinction between liberal and mechanical arts: only professions which do not require manual work were considered by them, their patrons, and their public to be worthy of the well-bred men.

The social antithesis of mechanical and liberal arts, of hands and tongue, influenced all intellectual and professional activity in the Renaissance. The uni-

versity-trained medical doctors contented themselves more or less with commenting on the medical writings of antiquity; the surgeons who did manual work
such as operating and dissecting belonged with the barbers and had a social
position similar to that of midwives. Literati were much more highly esteemed
than were artists. In the fourteenth century the latter were not separated from
whitewashers and stone-dressers and, like all craftsmen, were organized in
guilds. They gradually became detached from handicraft, until a separation was
effected in Italy about the end of the sixteenth century. In the period of
Leonardo da Vinci (about 1500) this had not yet been accomplished. This fact
appears rather distinctly in the writings of contemporary artists who over and
over again discussed the question as to whether painting and sculpture belong
with liberal or mechanical arts. In these discussions the painters usually stressed
their relations to learning (painting needs perspective and geometry) in order
to gain social esteem. Technological inventors and geographical discoverers, being
craftsmen and seamen, were hardly mentioned by the humanistic literati. The
great majority of the humanists did not report on them at all. If they mentioned
them, they did so in an exceedingly careless and inaccurate way. From the
present point of view the culture of the Renaissance owes its most important
achievements to the artists, the inventors, and the discoverers. Yet these men
entirely recede into the background in the literature of the period.

Beneath both the university scholars and the humanistic literati the artisans,
the mariners, shipbuilders, carpenters, foundrymen, and miners worked in silence
on the advance of technology and modern society. They had invented the
mariner's compass and guns; they constructed paper mills, wire mills, and stamping mills; they created blast furnaces and in the sixteenth century introduced
machines into mining. Having outgrown the constraints of guild tradition and
being stimulated to inventions by economic competition, they were, no doubt,
the real pioneers of empirical observation, experimentation, and causal research.
They were uneducated, probably often illiterate, and, perhaps for that reason,
today we do not even know their names. Among them were a few groups which
needed more knowledge for their work than their colleagues did and, therefore,
got a better education. Among these superior craftsmen the artists are most
important. There were no sharp divisions between painters, sculptors, goldsmiths,
and architects; but very often the same artist worked in several fields, since, on
the whole, division of labor had developed only slightly in the Renaissance.
Following from this a remarkable professional group arose during the fifteenth
century. The men we have in mind may be called artist-engineers, for not only
did they paint pictures, cast statues, and build cathedrals, but they also constructed lifting engines, canals and sluices, guns and fortresses. They invented new
pigments, detected the geometrical laws of perspective, and constructed new
measuring tools for engineering and gunnery. The first of them is Brunelleschi
(1377-1446), the constructor of the cupola of the cathedral of Florence.
Among his followers were Ghiberti (1377-1466), Leone Battista Alberti (1407-
72), Leonardo da Vinci (1492-1519), and Vanoccio Biringucci (d. 1538) whose
booklet on metallurgy is one of the first chemical treatises free of alchemistic
superstition. One of the last of them is Benvenuto Cellini (1500-1571), who was

a goldsmith and sculptor and also worked as military engineer of Florence. The German painter and engraver Albrecht Dürer, who wrote treatises on descriptive geometry and fortifications (1525 and 1527), belongs to this group. Many of the artist-engineers wrote—in the vernacular and for their colleagues—diaries and papers on their achievements. For the most part these papers circulated as manuscripts only. The artist-engineers got their education as apprentices in the workshops of their masters. Only Alberti had a humanistic education.

The surgeons belonged to a second group of superior artisans. Some Italian surgeons had contacts with artists, resulting from the fact that painting needs anatomical knowledge. The artificers of musical instruments were related to the artist-engineers. Cellini's father, for example, was an instrument-maker, and he himself was appointed as a pope's court musician for a time. In the fifteenth and sixteenth centuries the forerunners of the modern piano were constructed by the representatives of this third group. The makers of nautical and astronomical instruments and of distance meters for surveying and gunnery formed a fourth group. They made compasses and astrolabes, cross-staffs, and quadrants and invented the declinometer and inclinometer in the sixteenth century. Their measuring-instruments are the forerunners of the modern physical apparatus. Some of these men were retired navigators or gunners. The surveyors and the navigators, finally, were also considered as representatives of the mechanical arts. They and the map-makers are more important for the development of measurement and observation than of experimentation.

These superior craftsmen made contacts with learned astronomers, medical doctors, and humanists. They were told by their learned friends of Archimedes, Euclid, and Vitruvius; their inventive spirit, however, originated in their own professional work. The surgeons and some artists dissected, the surveyors and navigators measured, the artist-engineers and instrument-makers were perfectly used to experimentation and measurement, and their quantitative thumb rules are the forerunners of the physical laws of modern science. The occult qualities and substantial forms of the scholastics, the verbosity of the humanists were of no use to them. All these superior artisans had already developed considerable theoretical knowledge in the fields of mechanics, acoustics, chemistry, metallurgy, descriptive geometry, and anatomy. But, since they had not learned how to proceed systematically, their achievements form a collection of isolated discoveries. Leonardo, for example, deals sometimes quite wrongly with mechanical problems which, as his diaries reveal, he himself had solved correctly years before. The superior craftsmen, therefore, cannot be called scientists themselves, but they were the immediate predecessors of science. Of course, they were not regarded as respectable scholars by contemporary public opinion. The two components of scientific method were still separated before 1600—methodical training of intellect was preserved for upper-class learned people, for university scholars, and for humanists; experimentation and observation were left to more or less plebeian workers.

The separation of liberal and mechanical arts manifested itself clearly in the literature of the period. Before 1550 respectable scholars did not care for the achievements of the nascent new world around them and wrote in Latin. On the other hand, after the end of the fifteenth century, a literature published by "mechanics" in Spanish, Portuguese, Italian, English, French, Dutch, and German had developed. It included numerous short treatises on navigation, vernacular mathematical textbooks, and dialogues dealing with commercial, technological, and gunnery problems (e.g., Étienne de la Roche, Tartaglia, Dürer, Ympyn), and various vernacular booklets on metallurgy, fortification, bookkeeping, descriptive geometry, compass-making, etc. In addition there were the unprinted but widely circulated papers of the Italian artist-engineers. These books were diligently read by the colleagues of their authors and by merchants. Many of these books, especially those on navigation, were frequently reprinted, but as a rule they were disregarded by respectable scholars. As long as this separation persisted, as long as scholars did not think of using the disdained methods of manual workers, science in the modern meaning of the word was impossible. About 1550, however, with the advance of technology, a few learned authors began to be interested in the mechanical arts, which had become economically so important, and composed Latin and vernacular works on the geographical discoveries, navigation and cartography, mining and metallurgy, surveying, mechanics, and gunnery. Eventually the social barrier between the two components of the scientific method broke down, and the methods of the superior craftsmen were adopted by academically trained scholars: real science was born. This was achieved about 1600 with William Gilbert (1544–1603), Galileo (1564–1642), and Francis Bacon (1561–1626).

William Gilbert, physician to Queen Elizabeth, published the first printed book composed by an academically trained scholar which was based entirely on laboratory experiment and his own observation (*De magnete* [1600]). Gilbert used and invented physical instruments but neither employed mathematics nor investigated physical laws. Like a modern experimentalist he is critically-minded. Aristotelism, belief in authority, and humanistic verbosity were vehemently attacked by him. His scientific method derives from foundrymen, miners, and navigators with whom he had personal contacts. His experimental devices and many other details were taken over from a vernacular booklet of the compass-maker Robert Norman, a retired mariner (1581).

Galileo's relations to technology, military engineering, and the artist-engineers are often underrated. When he studied medicine at the University of Pisa in the eighties of the sixteenth century, mathematics was not taught there. He studied mathematics privately with Ostilio Ricci, who had been a teacher at the Accademia del Disegno in Florence, a school founded about twenty years earlier for young artists and artist-engineers. Its founder was the painter Vasari. Both the foundation of this school (1562) and the origin of Galileo's mathematical education show how engineering and its methods gradually rose from the workshops of craftsmen and eventually penetrated the field of academic instruction. As a young professor at Padua (1592–1610), Galileo lectured at the university on mathematics and astronomy and privately on mechanics and engineering. At this time he established workrooms in his house, where craftsmen were his assistants.

This was the first "university" laboratory in history. He started his research with studies on pumps, on the regulation of rivers, and on the construction of fortresses. His first printed publication (1606) described a measuring tool for military purposes which he had invented. All his life he liked to visit dockyards and to talk with the workmen. In his chief work of 1638, the *Discorsi*, the setting of the dialogue is the Arsenal of Venice. His greatest achievement—the detection of the law of falling bodies, published in the *Discorsi*—developed from a problem of contemporary gunnery, as he himself declared. The shape of the curve of projection had often been discussed by the gunners of the period. Tartaglia had not been able to answer the question correctly. Galileo, after having dealt with the problem for forty years, found the solution by combining crafts-man-like experimentation and measurement with learned mathematical analysis. The different social origin of the two components of his method—which became the method of modern science—is obvious in the *Discorsi*, since he gives the mathematical deductions in Latin and discusses the experiments in Italian. After 1610 Galileo gave up writing Latin treatises and addressed himself to nonscholars. His greatest works, consequently, are written completely or partially in Italian. A few vernacular poets were among his literary favorites. Even his literary taste reveals his predilection for the plain people. His aversion to the spirit and methods of the contemporary professors and humanists is frequently expressed in his treatises and letters.

The same opposition to both humanism and scholasticism can be found in the works of Francis Bacon. No scholar before him had attacked belief in authority and imitation of antiquity so passionately. Bacon was enthusiastic about the great navigators, the inventors, and the craftsmen of his period; their achievements, and only theirs, are set by him as models for scholars. The common belief that it is "a kind of dishonor to descend to inquiry upon matters mechanical"[1] seems "childish" to him. Induction, which is proclaimed by him as the new method of science, obviously is the method of just those manual laborers. He died from a cold which he caught when stuffing a chicken with snow. This incident also reveals how much he defied all customs of contemporary scholarship. An experiment of this kind was in his period considered worthy rather of a cook or knacker than of a former lord chancellor of England. Bacon, however, did not make any important discovery in the field of natural science, and his writings abound with humanistic rhetoric, scholastic survivals, and scientific mistakes. He is the first writer in the history of mankind, however, to realize fully the basic importance of methodical scientific research for the advancement of human civilization.

Bacon's real contribution to the development of science appears when he is confronted with the humanists. The humanists did not live on the returns from their writings but were dependent economically on bankers, noblemen, and princes. There was a kind of symbiosis between them and their patrons. The humanist received his living from his patron and, in return, made his patron famous by his writings. Of course, the more impressive the writings of the humanist, the more famous he became. Individual fame, therefore, was the professional ideal of the humanistic literati. They often called themselves "dispensers of glory" and quite openly declared fame to be the motive of their own and every intellectual activity. Bacon, on the contrary, was opposed to the ideal of individual glory. He substituted two new aims: "control of nature" by means of science and "advancement of learning." Progress instead of fame means the substitution of a personal ideal by an objective one. In his *Nova Atlantis* Bacon depicted an ideal state in which technological and scientific progress is reached by planned cooperation of scientists, each of whom uses and continues the investigations of his predecessors and fellow-workers. These scientists are the rulers of the New Atlantis. They form a staff of public officials organized in nine groups according to the principle of division of labor. Bacon's ideal of scientific cooperation obviously originated in the ranks of manufacturers and artisans. On the one hand, early capitalistic manual workers were quite accustomed to use the experience of their colleagues and predecessors, as is stressed by Bacon himself and occasionally mentioned by Galileo. On the other hand, division of labor had advanced in contemporary society and in the economy as a whole.

Essential to modern science is the idea that scientists must cooperate in order to bring about the progress of civilization. Neither disputing scholastics nor literati, greedy of glory, are scientists. Bacon's idea is substantially new and occurs neither in antiquity nor in the Renaissance. Somewhat similar ideas were pointed out in the same period by Campanella and, occasionally, by Stevin and Descartes. As is generally known, Bacon's *Nova Atlantis* greatly influenced the foundation of learned societies. In 1654 the Royal Society was founded in London, in 1663 the Académie française in Paris; in 1664 the *Proceedings* of the Royal Society appeared for the first time. Since this period cooperation of scientists in scientific periodicals, societies, institutes, and organizations has steadily advanced.

On the whole, the rise of the methods of the manual workers to the ranks of academically trained scholars at the end of the sixteenth century is the decisive event in the genesis of science. The upper stratum could contribute logical training, learning, and theoretical interest; the lower stratum added causal spirit, experimentation, measurement, quantitative rules of operation, disregard of school authority, and objective cooperation.

The indicated explanation of the development of science obviously is incomplete. Money economy and coexistent strata of skilled artisans and secular scholars are frequent phenomena in history. Why, nevertheless, did science not develop more frequently? A comparison with classical antiquity can fill at least one gap in our explanation.

Classical culture produced achievements in literature, art, and philosophy which are in no way inferior to modern ones. It produced outstanding and numerous historiographers, philologists, and grammarians. Ancient rhetoric is superior to its modern counterpart both in refinement and in the number of representatives. Ancient achievements are considerable in the fields of theoretical astronomy and mathematics, limited in the biological field, and poor in the physical sciences. Only three physical laws were correctly known to the ancient scholars: the principles of the lever and of Archimedes and the optical law of reflection. In the field of technology one difference is most striking: machines were used in antiquity in warfare, for juggleries, and for toys but were not employed in the production of goods. On the whole, ancient culture was borne by a rather small upper class living on their rents. Earning money by professional labor was always rather looked down upon in the circles determining ancient public opinion. Manual work was even less appreciated. In the same manner as in the Renaissance, painters and sculptors gradually detached from handicraft and slowly rose to social esteem. Yet their prestige never equaled that of writers and rhetors, and even in the period of Plutarch and Lucianus the greatest sculptors of antiquity would be attacked as manual workers and wage-earners. Compared with poets and philosophers, artists were rarely mentioned in literature, and engineers and technological inventors virtually never. The latter presumably (very little is known of them) were superior artisans or emancipated slaves working as foremen. In antiquity rough manual work was done by slaves.

As far as our problem is concerned, this is the decisive difference between classical and early capitalistic society. Machinery and science cannot develop in a civilization based on slave labor. Slaves generally are unskilled and cannot be intrusted with handling complex devices. Moreover, slave labor seems to be cheap enough to make introduction of machines superfluous. On the other hand, slavery makes the social contempt for manual work so strong that it cannot be overcome by the educated. For this reason ancient intellectual development could not overcome the barrier between tongue and hand. In antiquity only the least prejudiced among the scholars ventured to experiment and to dissect. Very few scholars, such as Hippocrates and his followers, Democritus, and Archimedes, investigated in the manner of modern experimental and causal science, and even Archimedes considered it necessary to apologize for constructing battering-machines. All these facts and correlations have already been pointed out several times.

It may be said that science could fully develop in modern Western civilization because European early capitalism was based on free labor. In early capitalistic society there were very few slaves, and they were not used in production but were luxury gifts in the possession of princes. Evidently lack of slave labor is a necessary but not a sufficient condition for the emergence of science. No doubt further necessary conditions would be found if early capitalistic society were compared with Chinese civilization. In China, slave labor was not predominant, and money economy had existed since about 500 B.C. Also there were in China, on the one hand, highly skilled artisans and, on the other, scholar-officials, approximately corresponding to the European humanists. Yet causal, experimental, and quantitative science not bound to authorities did not arise. Why this did not happen is as little explained as why capitalism did not develop in China.

The rise of science is usually studied by historians who are primarily interested in the temporal succession of the scientific discoveries. Yet the genesis of science can be studied also as a sociological phenomenon. The occupations of the scientific authors and of their predecessors can be ascertained. The sociological function of these occupations and their professional ideals can be analyzed. The temporal succession can be interrupted and relevant sociological groups

can be compared to analogous groups in other periods and other civilizations—
the medieval scholastics with Indian priest-scholars, the Renaissance humanists
with Chinese mandarins, the Renaissance artisans and artists with their colleagues
in classical antiquity. Since, in the sociology of culture, experiments are not
feasible, comparison of analogous phenomena is virtually the only way of finding
and verifying causal explanations. It is strange how rarely investigations of this
kind are made. As the complex intellectual constructs are usually studied
historically only, so sociological research for the most part restricts itself to
comparatively elementary phenomena. Yet there is no reason why the most
important and interesting intellectual phenomena should not be investigated
sociologically and causally.

Bibliography

1. *Novum Organum,* I, aph. 120.

BORIS HESSEN

The Social and Economic Roots of Newton's *Principia*

Boris Hessen, a young Soviet physicist, had a remarkable influence on the
history of science through a paper from which the following selections were
taken. Although often described as crude and exaggerated in its interpretation,
this paper is nevertheless cited as a breakthrough in the social history of science.
Hessen was unable to continue and refine his original ideas. He vanished in the
tragic Stalinist trials of the 1930's. Hessen's paper, originally presented at a
London conference on the history of science and technology, was one of several
by Soviet Russian scientists that provoked unusual debate, discussion, and
interest in the historical sociology of science. Using a Marxist conception of
historical explanation, Hessen attempts to show that economic factors were
decisively important in the development of modern physics. He does this by
showing that the problems and solutions of Newton's *Principia* were conditioned
by quite specific requirements of industrial and economic development.

Ways of Communication

By the beginning of the middle ages trade had already achieved considerable
development. Nevertheless, the land ways of communication were in a very miserable state. The roads were so narrow that even two horses could not pass. The
ideal road was one on which three horses could travel side-by-side, where, in the

The Social and Economic Roots of Newton's Principia by Boris Hessen. New York: Howard
Fertig, Inc., 1971, pp. 7–16, 24–26.

expression of the time (14th century) "A bride could ride by without touching the funeral cart."

Commonly, commodities were carried in packs. Road construction was almost nonexistent. The self-centered nature of feudal economy gave no impulse whatever to the development of road construction. On the contrary, both the feudal barons and the inhabitants of places through which commercial transport passed were interested in maintaining the poor condition of the roads, because they had the right of ownership to anything which fell onto their land from the cart or pack.

The speed of land transport in the fourteenth century did not exceed five to seven miles in the day.

Naturally maritime and water transport played a great part, both in consequence of the great load-capacity of the vessels and also of the greater speed of transit: the largest of two-wheeled carts drawn by ten to twelve oxen hardly carried two tons of goods, whereas an average sized vessel carried upwards of 600 tons. During the fourteenth century the journey from Constantinople to Venice took three times as long by land as by sea.

Nevertheless even the sea transport of this period was very imperfect: as sound methods of establishing the ship's position in the open sea had not yet been invented, they sailed close to the shores, which greatly retarded the speed of transit.

Although the first mention of the mariner's compass in the Arabian book "The Merchant's Treasury" dates to 1242, it came into universal use not earlier than the second half of the sixteenth century. Geographical maritime maps made their appearance about the same time.

But the compass and charts can be rationally exploited only when there is knowledge of methods of establishing the ship's position, i.e., when the latitude and longitude can be determined.

The development of merchant capital broke down the isolation of the town and the village commune, extended the geographical horizon to an extraordinary extent, and considerably accelerated the tempo of existence. It had need of convenient ways of communication, more perfect means of communication, a more exact measurement of time, especially in connection with the continually accelerating rate of exchange, and exact application of accounting and measuring.

Particular attention was directed to water transport: to maritime transport as a means of linking up various countries and to river transport as an internal link.

The development of river transport was also assisted by the fact that in antiquity waterways were the most convenient and most investigated, and the natural growth of the towns was linked up with the system of river communications. Transport over the rivers was three times as cheap as haulage transport.

The construction of canals also developed as a complementary means of internal transport and in order to link up the maritime transport with the internal river system.

Thus the development of merchant capital set transport the following technical problems:

In the Realm of Water Transport

1. An increase in the tonnage capacity of vessels and in their speed.
2. An improvement in the vessels' floating qualities: their reliability, sea-worthiness, their lesser tendency to rock, response to direction and ease of maneuvering, which was especially important for war vessels.
3. Convenient and reliable means of determining position at sea. Means of determining the latitude and longitude, magnetic deviation, times of tides.
4. The perfecting of the internal waterways and their linking up with the sea; the construction of canals and locks.

Let us consider what physical prerequisites are necessary in order to resolve these technical problems.

1. In order to increase the tonnage capacity of vessels it is necessary to know the fundamental laws governing bodies floating in liquids, since in order to estimate tonnage capacity it is necessary to know the method of estimating a vessel's water displacement. These are problems of hydrostatics.
2. In order to improve the floating qualities of a vessel it is necessary to know the laws governing the movement of bodies in liquids—this is an aspect of the laws governing the movement of bodies in a resistant medium—one of the basic tasks of hydrodynamics.
   The problem of a vessel's stability when rocking is one of the basic tasks of the mechanics of material points.
3. The problem of determining the latitude consists in the observation of heavenly bodies and its solution depends on the existence of optical instruments and a knowledge of the chart of the heavenly bodies and of their movement –of the mechanics of the heavens.
   The problem of determining longitude can be most conveniently and simply solved with the aid of a chronometer. But as the chronometer was invented only in the thirties of the eighteenth century after the work of Huygens, in order to determine the longitude recourse was made to measurement of the distance between the moon and the fixed stars.
   This method, put forward in 1498 by Amerigo Vespucci, demands an exact knowledge of the anomalies in the moon's movement and constitutes one of the most complicated tasks of the mechanics of the heavens. The determination of the times of the tides in dependence on the locality and on the position of the moon demands a knowledge of the theory of attraction, which also is a task of mechanics.
   How important this task was is evident from the circumstance that long before Newton gave the world his general theory of tides on the basis of the theory of gravity; in 1590, Stevin drew up tables in which was shown the time of the tides in any given place in dependence on the position of the moon.
4. The construction of canals and locks demands a knowledge of the basic laws of hydrostatics, the laws governing the efflux of liquids, since it is necessary to

know how to estimate the pressure of water and the speed of its efflux. In 1598 Stevin was occupied with the problem of the pressure of water and he saw that water could exert a pressure on the bottom of a vessel greater than its weight; in 1642 Castelli published a special treatise on the movement of water in canals of various sections. In 1646 Torricelli was working on the theory of efflux of fluids.

As we see, the problems of canal and lock construction also bring us to the tasks of mechanics (hydrostatics and hydrodynamics).

Industry

Already by the end of the middle ages (14th and 15th centuries) the mining industry was developing into a large industry. The mining of gold and silver in connection with the development of currency circulation was stimulated by the growth of exchange. The discovery of America was chiefly due to the gold famine, since European industry, which had developed so powerfully during the 14th and 15th centuries, and correspondingly European commerce, demanded larger supplies of the means of exchange; on the other hand the need for gold forced especial attention to be turned to the exploitation of mines and other sources of gold and silver.

The powerful development of the war industry, which had made enormous advances from the time of the invention of firearms and the introduction of heavy artillery, stimulated the exploitation of iron and copper mines to a tremendous extent. By 1350 firearms had become the customary weapon of the armies of eastern, southern and central Europe.

In the fifteenth century heavy artillery had reached a high level of perfection. In the 16th and 17th centuries the war industry made enormous demands upon the metallurgical industry. In the months of March and April 1652 alone, Cromwell required 335 cannon, and in December a further 1,500 guns of an aggregate weight of 2,230 tons, with 117,000 balls and 5,000 hand bombs in addition.

Consequently it is clear why the problem of the most effective exploitation of mines became a matter of prime importance.

First and foremost arises the problem set by the depth at which the ores lie. But the deeper the mines, the more difficult and dangerous work in them becomes.

A quantity of equipment for the pumping of water, the ventilation of the mines, and the raising of the ore to the surface becomes necessary. In addition a knowledge of the sound opening up of mines and of the plan of their workings is necessary.

By the beginning of the 16th century mining had reached a considerable development. Agricola left a detailed encyclopedia of mining from which one can see how much technical equipment had come to be applied in mining.

In order to raise the ore and to pump out water, pumps and lifting equipment (windlasses and horizontal worms) were constructed; the energy of animals, the wind and falling water were all put into service. A complete pumping system began to exist, since with the deepening of the mines the problem of removing the water becomes one of the most important of the technical tasks.

In his book Agricola describes three kinds of instruments for drawing away water, seven kinds of pumps, and six kinds of equipment for drawing off water by ladling or bucketing, altogether sixteen kinds of water-raising machines.

The development of mining involved enormous equipment for the working up of the ore. Here we meet with smelting furnaces, stamping mills, and machinery for dividing metals.

By the 16th century the mining industry had become a complex organism demanding considerable knowledge in its organization and direction. Consequently the mining industry at once develops as a large-scale industry, free of the craft system, and so not subject to craft stagnation. It was technically the most progressive and engendered the most revolutionary elements of the working class during the middle ages, i.e., the miners.

The cutting of galleries demands considerable knowledge of geometry and trigonometry. By the 15th century scientific engineers were working in the mines.

Thus the development of exchange and of the war industry set the mining industry the following technical problems:

1. The raising of ores from considerable depths.
2. Methods of ventilating the mines.
3. The pumping out of water and water-conducting equipment, the problem of the pump.
4. The transfer from the crude, damp-blast method of production predominant until the 15th century, to the more perfect form of blast-furnace production, in which the problem of air-blast equipment is raised, as it is in ventilation also.
5. The working up of the ores with the aid of rolling and cutting machinery.

Let us consider the problems of physics lying at the bases of these technical tasks.

1. The raising of ore and the task of equipping the raising machinery is a matter of arrangement of windlasses and blocks, i.e., of a variety of simple mechanical machines.
2. Ventilation equipment demands a study of draughts, i.e., it is a matter of aerostatics, which in turn is part of the task of statics.
3. The pumping of water from the mines and the equipment of pumps, especially of piston pumps, necessitates considerable investigation in the realm of hydro- and aerostatics.

    Consequently Torricelli, Herique, and Pascal occupied themselves with the problem of raising liquids in tubes and with atmospheric pressure.
4. The transfer to the blast-furnace production at once evoked the phenomenon of great blast-furnaces with the necessary buildings, water-wheels, bellows, rolling machines and heavy hammers.

    The problems of hydrostatics and dynamics set by the erection of water-wheels, the problem of air-bellows as also that of forced air for ventilation purposes also demand a study of the movement of air and its compression.

5. As in the case of other equipment, the construction of presses and heavy hammers brought into motion by utilizing the force of falling water (or animal power) demands a complicated planning of cogged wheels and transmission mechanism, which also is essentially a task of mechanics. In the mill develops the science of friction and the mathematical arrangement of cogged transmission wheels.

Thus, leaving out of account the great demands which the mining and metal-working industries of this period made on chemistry, all the aggregate of tasks of physics fell within the limits of mechanics.

War and War Industry

The history of war, Marx wrote to Engels in 1857, allows us more and more clearly to confirm the accuracy of our views on the connections between productive forces and social relationships.

Altogether the army is very important to economic development. It was in warfare that the craft order of corporations of artisans first originated. Here also we first find the application of machinery on a large scale.

Even the special value of metals and their role as currency were evidently based on their war significance.

So also the division of labor within various spheres of industry was first introduced in the army. Here in a tabloid form we find the entire history of the bourgeois system.

From the time of the application of gunpowder in Europe (it was used in China even before our era), a swift increase of firearms sets in.

Heavy artillery first appeared in 1280, during the siege of Cordova by the Arabs. In the 14th century firearms passed from the Arabs to the Spaniards. In 1308 Ferdinand IV took Gibraltar with the aid of cannon.

The first heavy guns were extremely unwieldy and they could only be transported in sections. Even weapons of small caliber were very heavy, since no proportion whatever had been established between the weight of the weapon and the ball and between the weight of the ball and the charge.

Nevertheless firearms were used not only in sieges, but on war-vessels. In 1386 the English captured two war-vessels armed with cannon.

A considerable improvement in artillery took place during the 15th century. Stone balls were replaced by iron. Cannon were cast solidly from iron and copper. Gun-carriages were improved and transport made great strides forward. The rate of fire was accelerated. To this factor is due the success of Charles VIII in Italy.

In the battle of Fornovo the French fired more shots in one hour than the Italians fired in a day.

Machiavelli wrote his "Art of War" specially in order to demonstrate means of resisting artillery by the artificial disposition of infantry and cavalry.

But of course the Italians were not satisfied with this alone, and they developed their own war industry. In Galileo's time the arsenal at Florence had attained to considerable development.

Francis I formed artillery into a separate unit and his artillery shattered the hitherto undefeated Swiss pikes.

The first theoretical works on ballistics and artillery date from the 16th century. In 1537 Tartaglia endeavored to determine the trajectory of the flight of a shot and established that the angle of 45 degrees allows the greatest distance to flight. He also drew up tables for directing aim.

Vanucci Biringuccio studied the process of casting and in 1540 he introduced considerable improvements in the production of weapons.

Hartmann invented a scale of calibers, by means of which each section of the gun could be measured in relation to the aperture, which gave a certain standard in the production of guns and opened the way for the introduction of fixed theoretical principles and empirical laws of firing.

In 1690 the first artillery school was opened in France.

In 1697 San-Remi published the first complete primer of artillery.

Towards the end of the 17th century in all countries artillery lost its medieval, craft character and was included as a component part of the army.

Consequently experiments on the inter-relationship of caliber and charge, the relationship of caliber to weight and length of barrel, on the phenomenon of recoil, developed on a large scale.

The progress of ballistics went hand in hand with the work of the most prominent of the physicists.

Galileo gave the world the theory of the parabolic trajectory of a ball; Torricelli, Newton, Bernouilli and Enler engaged in the investigation of the flight of a ball through the air, studied the resistance of the air and the causes of declination.

The development of artillery led in turn to a revolution in the construction of fortifications and fortresses, and this made enormous demands upon the engineering art.

The new form of defensive works (earthwork, fortresses) almost paralyzed the activity of artillery in the middle of the 17th century, and this in turn gave a mighty impulse to its further development.

The development of the art of war raised the following technical problems:

Intrinsic Ballistics

1. Study of the processes which occur in a firearm when fired and their improvement.
2. The stability combined with least weight of the firearm.
3. Adaptation to suitable and good aim.

Extrinsic Ballistics

4. The trajectory of a ball through a vacuum.
5. The trajectory of a ball through the air.
6. The dependence of air resistance upon the flight of the ball.
7. The deviation of a ball from its trajectory.

The Physical Bases of These Problems

1. Study of the processes which occur in the firearm demands study of the compression and extension of gases—in its basis a task of mechanics, and also study of the phenomenon of recoil (the law of action and counter-action).
2. The stability of a firearm raises the problem of studying the resistance of materials and of testing their durability. This problem, which also has great importance for the art of construction in the given stage of development, is resolved by purely mechanical means. Galileo gives considerable attention to the problem in his "Mathematical Demonstrations."
3. The problem of a ball's trajectory through a vacuum consists in resolving the task of the free fall of a body under the influence of gravity and the conjuncture of its progressive movement with its free fall. Naturally, therefore, Galileo gave much attention to the problem of the free fall of bodies. How far his work was connected with the interests of the artillery and ballistics can be judged if only from the fact that he begins his "Mathematical Demonstrations" with an address to the Florentines, in which he praises the activity of the arsenal at Florence and points out that the work of this arsenal provides a rich material for the scientific study.
4. The flight of a ball through the air is part of the problem of the movement of bodies through a resistant medium and of the dependence of that resistance upon the speed of the movement.
5. The deviation of the ball from the estimated trajectory can occur in consequence of a change in the initial speed of the ball, a change in the density of the atmosphere, or through the influence of the rotation of the earth. All these are purely mechanical problems.
6. Accurate tables governing aim can be drawn up provided the problem of extrinsic ballistics is resolved and the general theory of a ball's trajectory through a resistant medium is given.

Thus we see that if the process of the actual production of the firearm and the ball, which is a problem of metallurgy, be left out of account, the chief problems raised by the artillery of this period were problems of mechanics.

Now let us systematically consider the problems of physics raised by the development of transport, industry and mining.

First and foremost we have to note that all of them are purely problems of mechanics.

We analyze in a very general way the basic themes of research in physics during the period in which merchant capital was becoming the predominant economic force and manufacture began to develop, i.e., the period from the beginning of the 16th to the second half of the 17th century.

We do not include Newton's works on physics, since they will be subjected to a special analysis. A comparison of the basic themes of physics enables us to determine the basic tendency of the interests of physics during the period immediately preceding Newton and contemporary with him.

1. The problem of simple machines, sloping surfaces and general problems of

statics were studied by: Leonardo da Vinci (end of 16th century); Ubaldi (1577); Galileo (1589-1609); Cardan (middle of 16th century); and Stevin (1587).

2. The free fall of bodies and the trajectory of thrown bodies were studied by: Tartaglia (thirties of the 16th century); Benedetti (1587); Piccolomini (1598); Galileo (1589-1609); Riccioli (1652); The Academy del Cimente (1649).

3. The laws of hydro- and aerostatics, and atmospheric pressure. The pump, the movement of bodies through a resistant medium: Stevin, at the end of the 16th and beginning of the 17th centuries, the engineer and inspector of the land and water equipment of Holland; Galileo, Torricelli (first quarter of 17th century); Pascal (1647-1653); Herique (1650-1663), engineer to the army of Gustavus Adolphus, the builder of bridges and canals; Robert Boyle (seventies of the seventeenth century); Academy del Cimente (1657-1673).

4. Problems of the mechanics of the heavens, the theory of tides. Kepler (1609); Galileo (1609-1616); Gassendi (1647); Wren (sixties of 17th century); Halley (seventies of 17th century); Robert Hooke.

The above specified problems embrace almost the whole sphere of physics.

If we compare this basic series of themes with the physical problems which we found when analyzing the technical demands of transport, means of communication, industry and war, it becomes quite clear that these problems of physics were fundamentally determined by these demands.

In fact the group of problems stated in the first paragraph constitute the physical problems relating to raising equipment and transmission mechanism important to the mining industry and the building art.

The second group of problems has fundamental significance for artillery and constitute the basic physical tasks of ballistics.

The third group of problems is of fundamental importance to the problems of pumping water from mines and of their ventilation, the smelting of ores, the building of canals and locks, intrinsic ballistics and calculating the form of vessels.

The fourth group is of enormous importance to navigation.

All these are fundamentally mechanical problems. This of course does not mean that during this period other aspects of the movement of matter did not occupy attention. During this period optics began to develop and the first observations on static electricity and magnetism were made.[1] Nevertheless both by their nature and by their specific importance these problems have quite a subsidiary significance, and by the extent of their investigation and mathematical development (with the exception of certain laws of geometrical optics, which were of considerable importance in the construction of optical instruments) lagged far behind mechanics.

So far as optics were concerned this science received its main impulse from those technical problems which were of importance first and foremost to marine navigation.[2]

Now let us turn to an analysis of the contents of Newton's "Principia" and consider in what interrelationships they stand with the themes of physical research of the period.

In the definitions and axioms or laws of motion are expounded the theoretical and methodological bases of mechanics.

In the first book is a detailed exposition of the general laws of motion under the influence of central forces. In this way Newton provides a preliminary completion of the work to establish the general principles of mechanics which Galileo had begun.

Newton's laws provide a general method for the resolution of the great majority of mechanical tasks.

The second book, devoted to the problem of the movement of bodies, treats of a number of problems connected with the complex of problems which we have already noted.

The first three sections of the second book are devoted to the problem of the movement of bodies in a resistant medium in relation to various cases of the dependence of resistance upon speed (lineal resistance, resistance proportional to the second degree of speed and resistance proportional to part of the first part of the second degree).

As we have above shown when analyzing the physical problems of ballistics, the development of which was connected with the development of heavy artillery, the tasks set and accomplished by Newton are of fundamental significance to extrinsic ballistics.

The fifth section of the second book is devoted to the fundamentals of hydrostatics and the problems of floating bodies. The same section considers the pressure of gases and liquids under pressure.

When analyzing the technical problems set by the construction of vessels, canals, water-pumping and ventilating equipment, we saw that the physical themes of these problems relate to the fundamentals of hydrostatics and aerostatics.

The sixth section deals with the problem of the movement of pendulums against resistance.

The laws governing the swing of mathematical and physical pendulums in a vacuum were found by Huygens in 1673 and applied by him to the construction of pendulum clocks.

We have seen from Newton's letter to Aston of what importance were pendulum clocks in determining longitude. The application of clocks in determining longitude led Huygens to the discovery of centrifugal force and the changes in acceleration of the force of gravity.

When the pendulum clocks brought by Riche from Paris to Caen in 1673 displayed a retarded movement Huygens was able at once to explain the phenomenon by the changes in acceleration of the force of gravity. The importance attached by Huygens himself to clocks is evident from the fact that his chief work is called: "On pendulum clocks."

Newton's works continue this course, and just as he passed from the mathematical case of the movement of bodies in a resistant medium with lineal resistance to the study of an actual case of movement, so he passed from the mathematical pendulum to an actual case of a pendulum's movement in a resistant medium.

The seventh section of the second book is devoted to the problem of movements of liquids and the resistance of a thrown body.

In it problems of hydrodynamics are considered, among them the problem of the efflux of liquids and the flow of water through tubes. As was above shown, all these problems are of cardinal importance in the construction and equipment of canals and locks and in planning water-pumping equipment.

In the same section the laws governing the fall of bodies through a resistant medium (water and air) are studied. As we know, these problems are of considerable importance in determining the trajectory of a thrown body and the trajectory of a shot.

The third book of the "Principia" is devoted to the "System of the World." It is devoted to the problems of the movements of planets, the movement of the moon and the anomalies of that movement, the acceleration of the force of gravity and its variations, in connection with the problem of the inequality of movement of chronometers in sea-voyages and the problem of tides.

As we have above indicated, until the invention of the chronometer the movement of the moon was of fundamental importance in determining longitude. Newton returned to this problem more than once (in 1691). The study of the laws of the moon's movement was of fundamental importance in compiling exact tables for determining longitude, and the English "Council of Longitude" instituted a high reward for work on the moon's movement.

In 1713 Parliament passed a special bill to stimulate investigations in the sphere of determining longitude. Newton was one of the eminent members of the Parliamentary commission.

As we have pointed out in analyzing the sixth section, the study of the movement of the pendulum, begun by Huygens, was of great importance to navigation; consequently in the third book Newton studies the problem of the second pendulum and subjects to analysis the movement of clocks during a number of ocean expeditions: that of Halley to St. Helena in 1677, Varenne's and de Hais's voyage to Martinique and Guadeloupe in 1682, Couple's journey to Lisbon, etc. in 1697, and a voyage to America in 1700.

When analyzing the causes of tides Newton subjects the height of flow tides in various ports and river mouths to analysis, and discusses the problem of the height of flows in dependence on the local situation of the port and the forms of the flow.

This rough outline of the contents of the "Principia" exhibits the complete coincidence of the physical thematics of the period, which arose out of the needs of economics and technique, with the main contents of the "Principia," which in the full sense of the word is a survey and systematic resolution of all the main group of physical problems. And as by their character all these problems were problems of mechanics, it is clear that Newton's chief work was a survey of the mechanics of the earth and the heavenly bodies.

Bibliography

1. Investigations into magnetism developed under the direct influence of the study of the deviation of the compass in the world's magnetic field, which had first been met with during the first distant sea expeditions. Gilbert gave much attention to the problems of the earth's magnetism.
2. During this period optics developed through study of the problem of the telescope.

THOMAS S. KUHN

The Structure of Scientific Revolutions

Thomas S. Kuhn, formerly on the faculty of the University of California, is now Professor of History of Science at Princeton University. He is a leading figure in the field of the history of concept formation in the sciences. This selection is from his provocative and widely discussed book, *The Structure of Scientific Revolutions* (1962). There are essentially two views of scientific progress. The first and more traditional view is that science "progresses" by a gradual accumulation of factual knowledge, which in turn leads to the expansion of theories to cover these newly discovered facts. The second is that major changes in science are "revolutionary"; formerly established facts and theories are not simply modified and expanded, but are abolished. In this second view, whole conceptual frameworks are overthrown and totally replaced in a relatively short time span. Kuhn, who is largely responsible for this second view, argues in its favor in this selection. Notice, in particular, his definitions of "normal science" and "paradigm" and how they influence scientific practice.

History, if viewed as a repository for more than anecdote or chronology, could produce a decisive transformation in the image of science by which we are now possessed. That image has previously been drawn, even by scientists themselves, mainly from the study of finished scientific achievements as these are recorded in the classics and, more recently, in the textbooks from which each new scientific generation learns to practice its trade. Inevitably, however, the aim of such books is persuasive and pedagogic; a concept of science drawn from them is no more likely to fit the enterprise that produced them than an image of a national culture drawn from a tourist brochure or a language text. This essay attempts to show that we have been misled by them in fundamental ways. Its aim is a sketch of the quite different concept of science that can emerge from the historical record of the research activity itself.

Even from history, however, that new concept will not be forthcoming if historical data continue to be sought and scrutinized mainly to answer questions posed by the unhistorical stereotype drawn from science texts. Those texts have, for example, often seemed to imply that the content of science is uniquely exemplified by the observations, laws, and theories described in their pages. Almost as regularly, the same books have been read as saying that scientific methods are simply the ones illustrated by the manipulative techniques used in gathering textbook data, together with the logical operations employed when relating those data to the textbook's theoretical generalizations. The result has been a concept of science with profound implications about its nature and development.

If science is the constellation of facts, theories, and methods collected in current texts, then scientists are the men who, successfully or not, have striven to contribute one or another element to that particular constellation. Scientific development becomes the piecemeal process by which these items have been added, singly and in combination, to the ever growing stockpile that constitutes scientific technique and knowledge. And history of science becomes the discipline that chronicles both these successive increments and the obstacles that have inhibited their accumulation. Concerned with scientific development, the historian then appears to have two main tasks. On the one hand, he must determine by what man and at what point in time each contemporary scientific fact, law, and theory was discovered or invented. On the other, he must describe and explain the congeries of error, myth, and superstition that have inhibited the more rapid accumulation of the constituents of the modern science text. Much research has been directed to these ends, and some still is.

In recent years, however, a few historians of science have been finding it more and more difficult to fulfil the functions that the concept of development-by-accumulation assigns to them. As chroniclers of an incremental process, they discover that additional research makes it harder, not easier, to answer questions like: When was oxygen discovered? Who first conceived of energy conservation? Increasingly, a few of them suspect that these are simply the wrong sorts of questions to ask. Perhaps science does not develop by the accumulation of individual discoveries and inventions. Simultaneously, these same historians confront growing difficulties in distinguishing the "scientific" component of past observation and belief from what their predecessors had readily labeled "error" and "superstition." The more carefully they study, say, Aristotelian dynamics, phlogistic chemistry, or caloric thermodynamics, the more certain they feel that those once current views of nature were, as a whole, neither less scientific nor more the product of human idiosyncrasy than those current today. If these out-of-date beliefs are to be called myths, then myths can be produced by the same sorts of methods and held for the same sorts of reasons that now lead to scientific knowledge. If, on the other hand, they are to be called science, then science has included bodies of belief quite incompatible with the ones we hold today. Given these alternatives, the historian must choose the latter. Out-of-date theories are not in principle unscientific because they have been discarded. That choice, however, makes it difficult to see scientific development as a process of accretion. The same historical research that displays the difficulties in isolating individual inventions and discoveries gives ground for profound doubts about the cumulative process through which these individual contributions to science were thought to have been compounded.

The result of all these doubts and difficulties is a historiographic revolution in the study of science, though one that is still in its early stages. Gradually, and often without entirely realizing they are doing so, historians of science have begun to ask new sorts of questions and to trace different, and often less than cumulative, developmental lines for the sciences. Rather than seeking the permanent contributions of an older science to our present vantage, they attempt to display the historical integrity of that science in its own time. They ask, for example, not about the relation of Galileo's views to those of modern science, but rather about the relationship between his views and those of his group, i.e., his teachers, contemporaries, and immediate successors in the sciences. Furthermore, they insist upon studying the opinions of that group and other similar ones from the viewpoint—usually very different from that of modern science—that gives those opinions the maximum internal coherence and the closest possible fit to nature. Seen through the works that result, works perhaps best exemplified in the writings of Alexandre Koyré, science does not seem altogether the same enterprise as the one discussed by writers in the older historiographic tradition. By implication, at least, these historical studies suggest the possibility of a new image of science. This essay aims to delineate that image by making explicit some of the new historiography's implications.

What aspect of science will emerge to prominence in the course of this effort? First, at least in order of presentation, is the insufficiency of methodological directives, by themselves, to dictate a unique substantive conclusion to many sorts of scientific questions. Instructed to examine electrical or chemical phenomena, the man who is ignorant of these fields but who knows what it is to be scientific may legitimately reach any one of a number of incompatible conclusions. Among those legitimate possibilities, the particular conclusions he does arrive at are probably determined by his prior experience in other fields, by the accidents of his investigation, and by his own individual makeup. What beliefs about the stars, for example, does he bring to the study of chemistry or electricity? Which of the many conceivable experiments relevant to the new field does he elect to perform first? And what aspects of the complex phenomenon that then results strike him as particularly relevant to an elucidation of the nature of chemical change or of electrical affinity? For the individual, at least, and sometimes for the scientific community as well, answers to questions like these are often essential determinants of scientific development. . . .

That element of arbitrariness does not, however, indicate that any scientific group could practice its trade without some set of received beliefs. Nor does it make less consequential the particular constellation to which the group, at a given time, is in fact committed. Effective research scarcely begins before a scientific community thinks it has acquired firm answers to questions like the following: What are the fundamental entities of which the universe is composed? How do these interact with each other and with the senses? What questions may legitimately be asked about such entities and what techniques employed in seeking solutions? At least in the mature sciences, answers (or full substitutes for answers) to questions like these are firmly embedded in the educational initiation that prepares and licenses the student for professional practice. Because that education is both rigorous and rigid, these answers come to exert a deep hold on the scientific mind. That they can do so does much to account both for the peculiar efficiency of the normal research activity and for the direction in which it proceeds at any given time. . . .

Yet that element of arbitrariness is present, and it too has an important effect on scientific development. . . . Normal science, the activity in which most scientists inevitably spend almost all their time, is predicated on the assumption that the scientific community knows what the world is like. Much of the success of the enterprise derives from the community's willingness to defend that assumption, if necessary at considerable cost. Normal science, for example, often suppresses fundamental novelties because they are necessarily subversive of its basic commitments. Nevertheless, so long as those commitments retain an element of the arbitrary, the very nature of normal research ensures that novelty shall not be suppressed for very long. Sometimes a normal problem, one that ought to be solvable by known rules and procedures, resists the reiterated onslaught of the ablest members of the group within whose competence it falls. On other occasions a piece of equipment designed and constructed for the purpose of normal research fails to perform in the anticipated manner, revealing an anomaly that cannot, despite repeated effort, be aligned with professional expectation. In these and other ways besides, normal science repeatedly goes astray. And when it does—when, that is, the profession can no longer evade anomalies that subvert the existing tradition of scientific practice—then begin the extraordinary investigations that lead the profession at last to a new set of commitments, a new basis for the practice of science. The extraordinary episodes in which that shift of professional commitments occurs are the ones known in this essay as scientific revolutions. They are the tradition-shattering complements to the tradition-bound activity of normal science.

The most obvious examples of scientific revolutions are those famous episodes in scientific development that have often been labeled revolutions before . . . associated with the names of Copernicus, Newton, Lavoisier, and Einstein. More clearly than most other episodes in the history of at least the physical sciences, these display what all scientific revolutions are about. Each of them necessitated the community's rejection of one time-honored scientific theory in favor of another incompatible with it. Each produced a consequent shift in the problems available for scientific scrutiny and in the standards by which the profession determined what should count as an admissible problem or as a legitimate problem-solution. And each transformed the scientific imagination in ways that we shall ultimately need to describe as a transformation of the world within which scientific work was done. Such changes, together with the controversies that almost always accompany them, are the defining characteristics of scientific revolutions.

These characteristics emerge with particular clarity from a study of, say, the Newtonian or the chemical revolution. It is, however, a fundamental thesis of this essay that they can also be retrieved from the study of many other episodes that were not so obviously revolutionary. For the far smaller professional group affected by them, Maxwell's equations were as revolutionary as Einstein's, and they were resisted accordingly. The invention of other new theories regularly, and appropriately, evokes the same response from some of the specialists on whose area of special competence they impinge. For these men the new theory implies a change in the rules governing the prior practice of normal science. Inevitably, therefore, it reflects upon much scientific work they have already successfully completed. That is why a new theory, however special its range of application, is seldom or never just an increment to what is already known. Its assimilation requires the reconstruction of prior theory and the re-evaluation of prior fact, an intrinsically revolutionary process that is seldom completed by a single man and never overnight. No wonder historians have had difficulty in dating precisely this extended process that their vocabulary impels them to view as an isolated event.

Nor are new inventions of theory the only scientific events that have revolutionary impact upon the specialists in whose domain they occur. The commitments that govern normal science specify not only what sorts of entities the universe does contain, but also, by implication, those that it does not. It follows, though the point will require extended discussion, that a discovery like that of oxygen or X-rays does not simply add one more item to the population of the scientist's world. Ultimately it has that effect, but not until the professional community has re-evaluated traditional experimental procedures, altered its conception of entities with which it has long been familiar, and, in the process, shifted the network of theory through which it deals with the world. Scientific fact and theory are not categorically separable, except perhaps within a single tradition of normal-scientific practice. That is why the unexpected discovery is not simply factual in its import and why the scientist's world is qualitatively transformed as well as quantitatively enriched by fundamental novelties of either fact or theory. . . .

Undoubtedly, some readers will already have wondered whether historical study can possibly effect the sort of conceptual transformation aimed at here. An entire arsenal of dichotomies is available to suggest that it cannot properly do so. History, we too often say, is a purely descriptive discipline. The theses suggested above are, however, often interpretive and sometimes normative. Again, many of my generalizations are about the sociology or social psychology of scientists; yet at least a few of my conclusions belong traditionally to logic or epistemology. . . .

Having been weaned intellectually on these distinctions and others like them, I could scarcely be more aware of their import and force. For many years I took them to be about the nature of knowledge, and I still suppose that, appropriately recast, they have something important to tell us. Yet my attempts to apply them, even *grosso modo,* to the actual situations in which knowledge is gained, accepted, and assimilated have made them seem extraordinarily problematic. Rather than being elementary logical or methodological distinctions, which would thus be prior to the analysis of scientific knowledge, they now seem integral parts of a traditional set of substantive answers to the very questions upon which they have been deployed. That circularity does not at all invalidate them. But it does make them parts of a theory and, by doing so, subjects them to the same scrutiny regularly applied to theories in other fields. If they are to have more than pure abstraction as their content, then that content must be discovered by observing them in application to the data they are meant to elucidate. How could history of science fail to be a source of phenomena to which theories about knowledge may legitimately be asked to apply?

In this essay, 'normal science' means research firmly based upon one or more past scientific achievements, achievements that some particular scientific community acknowledges for a time as supplying the foundation for its further practice. Today such achievements are recounted, though seldom in their original form, by science textbooks, elementary and advanced. These textbooks expound the body of accepted theory, illustrate many or all of its successful applications, and compare these applications with exemplary observations and experiments. Before such books became popular early in the nineteenth century (and until even more recently in the newly matured sciences), many of the famous classics of science fulfilled a similar function. Aristotle's *Physica,* Ptolemy's *Almagest,* Newton's *Principia* and *Opticks,* Franklin's *Electricity,* Lavoisier's *Chemistry,* and Lyell's *Geology*—these and many other works served for a time implicitly to define the legitimate problems and methods of a research field for succeeding generations of practitioners. They were able to do so because they shared two essential characteristics. Their achievement was sufficiently unprecedented to attract an enduring group of adherents away from competing modes of scientific activity. Simultaneously, it was sufficiently open-ended to leave all sorts of problems for the redefined group of practitioners to resolve.

Achievements that share these two characteristics I shall henceforth refer to as 'paradigms,' a term that relates closely to 'normal science.' By choosing it, I mean to suggest that some accepted examples of actual scientific practice— examples which include law, theory, application, and instrumentation together— provide models from which spring particular coherent traditions of scientific research. These are the traditions which the historian describes under such rubrics as 'Ptolemaic astronomy' (or 'Copernican'), 'Aristotelian dynamics' (or 'Newtonian'), 'corpuscular optics' (or 'wave optics'), and so on. The study of paradigms, including many that are far more specialized than those named illustratively above, is what mainly prepares the student for membership in the particular scientific community with which he will later practice. Because he there joins men who learned the bases of their field from the same concrete models, his subsequent practice will seldom evoke overt disagreement over fundamentals. Men whose research is based on shared paradigms are committed to the same rules and standards for scientific practice. That commitment and the apparent consensus it produces are prerequisites for normal science, i.e., for the genesis and continuation of a particular research tradition. . . .

If the historian traces the scientific knowledge of any selected group of related phenomena backward in time, he is likely to encounter some minor variant of a pattern here illustrated from the history of physical optics. Today's physics textbooks tell the student that light is photons, i.e., quantum-mechanical entities that exhibit some characteristics of waves and some of particles. Research proceeds accordingly, or rather according to the more elaborate and mathematical characterization from which this usual verbalization is derived. That characterization of light is, however, scarcely half a century old. Before it was developed by Planck, Einstein, and others early in this century, physics texts taught that light was transverse wave motion, a conception rooted in a paradigm that derived ultimately from the optical writings of Young and Fresnel in the early nineteenth century. Nor was the wave theory the first to be embraced by almost all practitioners of optical science. During the eighteenth century the paradigm for this field was provided by Newton's *Opticks*, which taught that light was material corpuscles. At that time physicists sought evidence, as the early wave theorists had not, of the pressure exerted by light particles impinging on solid bodies.

These transformations of the paradigms of physical optics are scientific revolutions, and the successive transition from one paradigm to another via revolution is the usual developmental pattern of mature science. . . .

STEPHEN E. TOULMIN

The Evolutionary Development of Natural Science

Stephen E. Toulmin was formerly a member of the philosophy faculties at Brandeis and Michigan State Universities. He has made significant contributions to the fields of ethics, theory of knowledge, and the history and philosophy of science. His most recent book is *Human Understanding* Vol. I (1971). He is presently Provost of Crown College, University of California at Santa Cruz. In this essay Toulmin argues against the view defended by Kuhn in the previous selection and in favor of an evolutionary interpretation of the history of science. Specifically, he rejects the idea of absolute scientific revolutions, suggesting that in the transition from one theoretical system to another there is never a complete break, nor is there mere replication. There are stops and starts, blind alleys and promising ones, and many varieties of evolutionary interaction and cross-fertilization.

In the course of the first three centuries of modern science—from around 1600 A.D. until a generation ago—all aspects of the natural world in turn came under the scientist's scrutiny: the stars and the earth, living creatures and their fossil remains, atoms and cells, chickadees and chimpanzees, primitive societies and mental disorders. I say "all aspects," but it would be more exact to say "nearly all." For, throughout this period, one thing was generally exempted from the scope of scientific inquiry: although with the passage of time many aspects of human behavior came to be studied from different points of view—so giving rise to the new sciences of ethnology, anthropology, sociology, and abnormal psychology—the activities of *the scientist himself* were not normally considered a suitable object for scientific study and analysis. Right through the nineteenth century, any suggestion of a "science of science" would have struck

Stephen E. Toulmin, "The Evolutionary Development of Natural Science." *American Scientist* 55:456–471, April 1967. Reprinted with the permission of the author and publisher.

men as a kind of *lèse-raison*. The business of science (it was thought) is to study the causes of natural phenomena; whereas science itself, as a rational activity, presumably operated on a higher level, and could not be thought of as a "natural phenomenon."

More recently, this self-denying ordinance has been somewhat relaxed. Twentieth-century science is less committed than the science of earlier centuries to explaining its phenomena in terms of rigid, mechanistic, cause-and-effect ideas. As a result, some of the restrictions earlier placed on scientific inquiries have been weakened, and the nature and working of science itself have been analyzed from various different points of view. Let me begin by reminding you about three of these lines of attack, which have up to now been followed largely independently.

1. To begin with, the development of natural science has been studied in a quantitative, statistical manner. For more than a century, since the pioneer work of Quetelet, statisticians have been developing techniques for describing and analyzing organic populations and growth-processes. As a result, it has become a commonplace that certain standard forms of growth-curve recur in a wide range of contexts, both biological and sociological: so that one and the same numerical pattern may be manifested equally in the growth of a bean-stalk, the spread of an infectious disease through a population, and the sales of domestic refrigerators. (The classic account of this general theory is to be found in D'Arcy Thompson's splendid treatise *On Growth and Form*.) Yet it was barely ten years ago that Professor Derek Price of Yale first demonstrated that these very same growth-patterns are discoverable also in the statistics of scientific activity.[1] If we provide ourselves with numerical indices for measuring the sheer quantity of scientific work being done at any time, we find (Price showed) those very same "S-shaped" or "logistic" growth-curves which are already familiar in the case of organic activities of other kinds.

The activities of scientists, accordingly, can be subjected to numerical analysis as legitimately—at any rate—as the activities of other social groups and professions. For what they are worth, the resulting discoveries can be highly suggestive. Not that they tell us everything, by any means: the answers to which social statistics can lead us are limited by the questions which statistical method permits us to ask. The same kind of limitation is involved here as (for instance) in gas theory, where thermodynamics and statistical mechanics give a great deal of insight at a macroscopic level, but tell us only the very minimum about the individual molecules of different gases. (Much of the virtue of statistical mechanics, indeed, lies in the fact that it is neutral as between different gases. So too, the sociometrics of science is, inevitably, neutral as between scientific inquiries of different kinds.) The content and merit of different pieces of scientific work must be judged, first, by criteria drawn from outside the statistics of science: in the nature of the case, sociometric methods of inquiry give us only numerical answers to quantitative questions.

2. Meanwhile, other scholars and scientists have been studying the development of science from a different, genetic point of view. Their concerns are with the internal development of the scientific tradition, and with the processes by which

scientific ideas grow out of and displace one another. For them, the process of scientific development is to be thought of, not so much as a quantitative, organic growth-process, but rather as a dialectical sequence: problems lead to solutions, which in turn lead to new problems, whose solutions pose new problems again . . . investigations yield ideas, which provide material for new investigations, out of which emerge further ideas . . . and so on.

This genetic or problematic approach to the development of science can be considered from two somewhat different points of view: sociohistorical or logico-philosophical. One may study the problematic development of science in the hope of building up an historical understanding of the characteristic processes of intellectual change in natural science; or alternatively one may aim at producing a logical analysis (or "rational reconstruction") of the methods of inquiry and argument by which scientific progress is properly made. Either way, this approach to the study of scientific development also is subject to a certain self-limitation. It gives an account of scientific development in which factors *outside* the disciplinary procedures of the natural science in question are referred to only marginally, if at all. To use a biological metaphor: it studies the ontogeny or morphogenesis of a science in isolation from its ecological environment. Clearly, for many purposes, the resulting abstraction may be both legitimate and fruitful; but it too is, nevertheless, an abstraction.

3. If the morphogenetic study of development abstracts a particular science from its wider environment, and considers its internal development in isolation, the natural complement consists in a purely sociological approach to the development of science. And many people during the last half century have indeed been drawn toward a study of the external, environmental interactions between science—regarded as a social phenomenon—and the larger culture or society within which the scientist has to operate: its institutions, its social structure, politics, and economics.

Once again, many of the results have been profoundly interesting, and in some cases unexpected. The work of such men as Dean Don K. Price of Harvard has led us to understand in a new way the manner in which the different scientific subdisciplines have become organized into institutional "guilds," and the processes by which scientific work has acquired the new economic, political, and social impact characteristic of the last hundred years.[2] Meanwhile, there has been a perennial temptation to look for a "feedback" from the social context into the actual content of scientific ideas: to speculate (for instance) that, in some manner or other, the development of thermodynamic theory in the first half of the nineteenth century *must* reflect in its structure contemporary developments in the technology of steam-locomotives. (Yet the actual *form* of this influence has up to now proved elusive.) Others have gone further, and hinted, e.g., the Darwin's theory of natural selection should be thought of as reflection of contemporary beliefs about *laissez-faire* economics—at which point, most readers begin to feel that ingenuity has lapsed into implausibility.[3] In such a generalized sociology of science, as in the numerical statistics of scientific growth, one must feel that the processes by which the content of science develops slip through our intellectual sieve; up to now attempts to force answers about *content* out of questions about the *social ecology* of science seem only to have distorted that content.

Yet it is worth asking: "Can we not find a fresh standpoint, from which we can preserve the real virtues of all these three distinct approaches, within the framework of a single, coherent account of scientific development?" If each of the three approaches does have real merit, it must surely be possible to harmonize them. For, manifestly, the internal development of scientific thought does have a kind of rationality and method; even though accident, spontaneity, and in some cases inspired blundering, have had their parts to play. Manifestly, too, there are quite genuine interactions between scientific thought and its social environment, although these are more subtle than the naïver Marxists would imply. The question, therefore, is: "How are we to bring these approaches closer together? Can we look at all these questions from a standpoint which makes the nature of their convergence more evident?"

This problem will be our chief topic in all that follows. The task will be to argue our way to a provisional model for analyzing the process of scientific development—a model, in the sense of a theoretical pattern showing the interrelations of different concepts and questions; but a merely provisional one, since on this occasion we can deal with the problem only to a first-order (perhaps even a "zeroth-order") approximation. Still, despite the crudity of this initial treatment, it will perhaps be a worthwhile achievement if we can simply establish that the three familiar approaches toward the study of scientific development *are* harmonizable within a larger, more integrated account.

1. To begin at the latter end of the spectrum: the social history of science has, as one of its central problems, the question, "What conditions must hold if there are to be any opportunities for scientific innovation at all?" Notice that this is not primarily a psychological question, since one may take it for granted that, within any population whatever, there will be minority of human beings having the necessary innate curiosity. Essentially, it is a sociological question, arising out of the observation that different societies and cultures, at different stages in their history, provide different opportunities and/or incentives to intellectual innovation or, more commonly, put different obstacles and/or disincentives in the way of intellectual heterodoxy. (If we ask, for instance, why cosmology and astronomy developed more slowly in China than in the West, we must bear in mind the tendency of eminent Chinese in the classical period to complain about the prevalence of unconventional ideas: where the cultural elite regards intellectual innovations as "dangerous thoughts," the institutions needed for the effective development of new scientific ideas can hardly flourish.[4]

Indeed, whenever one turns to consider the development of science in any particular culture, nation, or epoch, one fruitful first question can be, "On whose back was Science riding at this stage?" Just because disinterested curiosity about the natural world, being in itself a "pure" form of intellectual activity, pays no particular dividends beyond the satisfactions of better understanding, it has never by itself given men a living. The fruitful development of science has always been contingent on other activities or institutions, which—inadvertently or by design—have provided occasions for men to pursue scientific investigations. In retrospect, it may be obvious that the development of natural science is one of the crucial achievements of human civilization; yet, sociologically speaking, scientific activities have hitherto been merely epiphenomena.

If men in earlier epochs and other cultures have "changed their minds" about Nature, this has happened always as a *by-product* of activities having more direct social, economic, or political functions. In the great days of Babylon, for instance, striking progress took place in computational astronomy; but the men concerned developed these techniques in their capacity as government servants— for purposes of official prognostication and calendrical computation. In medieval Islam, again, the natural sciences of Greek antiquity were kept alive, and developed further, at a time when they were languishing in Europe; but there, too, the men responsible earned their living by other means—in most cases, as court physicians. Among the men who established the Royal Society in seventeenth-century London, a few were scholars of independent means, yet many of them needed other sources of professional income; and the finance for the Royal Society itself was obtained from King Charles II by the Secretary of the Admiralty (Samuel Pepys, the diarist, who was also the first Secretary of the Royal Society), through the same concatenation of circumstances that led so much American research in the 1950's to be financed through the Office of Naval Research. In the next century, we find the Anglican and Dissenting Churches providing employment for educated men which left them enough surplus energy and resources to pursue significant scientific work as well . . . And so the story goes on; with the National Aeronautics and Space Administration as only one more in a long sequence of institutions which have provided extraneous occasions for the scientifically-minded to exercise their disinterested curiosities.

To sum up this first group of questions: what opportunities any culture provides for the development of heterodox ideas about nature, and what *volume of innovation* one finds there, are matters which depend predominantly on factors *external* to the scientific developments in question. Faced with problems concerning the volume of scientific work being done on a given subject within some particular society, we can reasonably enough cite social, economic, institutional, and similar factors as the major considerations bearing on such issues. Even here, one has to qualify the generalization by the use of such words as "predominant" and "major," for the "ripeness" or "unripeness" of a particular subject also serves to enhance or inhibit intellectual curiosity. (When a problem shows signs of yielding to investigation, a bandwagon effect frequently follows; and conversely, a recalcitrant field of inquiry will remain comparatively neglected despite otherwise favorable social and institutional conditions; but these qualifications are—arguably—second-order ones.)

2. So much for the factors which determine how large a pool of scientific variants and novelties is under consideration at any particular place and time. But, when we turn our attention away from the sheer *size* of this pool, and start to ask questions about its *contents*, the picture begins to change. For why (we may ask) do scientists choose the particular new lines of thought (innovations, variants) they do? What considerations incline them to favor—say—"corpuscular,"

"fluid," or "field" theories of physical phenomena at any particular stage, and to ignore alternative possibilities, even when experiment does not choose between them? Where a dominant *direction* of variation can be observed within any particular science, or where some particular direction of innovation appears to have been excessively neglected, a new type of issue arises. Within the total volume of intellectual variants under discussion, what factors determine which types of option are, and which are not pursued?

Questions of this kind are, perhaps, the most complicated that can arise for the historian of scientific thought. On the one hand, the considerations which incline scientists working in neurophysiology (say) or atomic physics or optics, at any particular stage, to take certain general types of hypotheses more seriously than others must undoubtedly be related to the intellectual situation within that branch of natural science at the moment in question. (Notice: we are here concerned with the "initial plausibility" attributed to certain classes of hypotheses, not with their "verification" or "establishment." We are asking how scientists come to take certain kinds of new suggestions seriously in the first place— considering them to be worthy of investigation at all—rather than with the standards they apply in deciding that those suggestions are in fact sound and acceptable.) So it is clear that the existence and continuity of certain "schools," "fashions," or "points of view" within, say, physical theory must be regarded essentially as an internal, professional matter; and will need to be analyzed and explained, substantially, in terms of the longer-term historical evolution of ideas within that particular area of science.

Even so, this will rarely be the whole story, and sometimes it may be only a small part of it. In plenty of cases, the justification for taking a particular kind of scientific hypothesis seriously has to be sought outside the intellectual content of that particular science. The influence of Platonist ideas on Johann Kepler, for instance, shows that any attempt to draw a hard and fast boundary around "astronomical" considerations would be in vain. Likewise in nineteenth-century zoology: there too, a satisfactory story must bring in, e.g., the inhibiting influence of orthodox natural theology, on the one hand, and the positive influence of Malthus' theories of human population-growth on the other. When we are concerned with the content of the pool of intellectual variants, accordingly, rather than with its sheer volume, we have to consider this as the product— in varying proportions—of both "internal" and "external" factors.

3. However, if we proceed still further along the spectrum of possible questions, we shall find the balance tilting sharply in the other direction. Consider the question: "What factors determine which of the intellectual variants circulating in any generation are selected out and incorporated into the tradition of scientific thought?" Evidently enough, the course of intellectual change within the sciences depends not merely on intellectual variation, but even more on the collective decisions by which certain new suggestions are generally accepted as "established" and transmitted to the next generations of scientists as "well-attested" results. The crucial factor in this selection-process is the set of criteria in the light of which that choice is made. How do scientists determine this choice? Faced with that question, we must give a double answer—in part, one

concerned with aspirations; in part, one which recognizes historical actualities.

Suppose we consider only aspirations: i.e., the explicit program to which natural scientists would subscribe as a question (so to speak) of ideology. As a matter of broad principle, scientists commonly take it for granted that their criteria of "truth," "verification," or "falsification" are stateable in absolute terms. In principle, that is, these criteria should be the same for scientists in all epochs, in all cultures, and should remain unaffected by such factors as political prejudice and theological conservatism. To formulate the criteria in explicit terms may be a taxing and contentious task, but at any rate (they believe) one is entitled to demand that any solution to this problem shall provide a satisfactory "demarcation criterion" for setting off irrelevant, "extrinsic" considerations from relevant, "intrinsic" considerations.

So much for theoretical aspirations; but, when we turn to look at historical actualities, the picture becomes slightly more complex. True, one may certainly argue that these selection-criteria are—and are rightly—determined *predominantly* by the professional values of the community of scientists in question. (This, as Michael Polanyi has argued, is one fundamental element in the political theory of the "Republic of Science").[5] Yet there are reasons for wondering whether, in actual fact, this absolute independence of the selection-criteria from social and historical factors *has* ever been entirely realized; or, indeed, whether it ever *could* be. Many people will recall the passage in Pierre Duhem's book, *The Aim and Structure of Physical Theory,* in which he compares and contrasts the *styles* of theory found acceptable, respectively, by physical scientists in nineteenth-century Britain and France. French physicists writing about electricity and magnetism (he points out) demanded formal, axiomatized, mathematical expositions, with all the assumptions and deductions set out clearly and unambiguously. British physicists working in the same area operated, rather, in terms of mechanical models: these were to a large extent intuitive rather than explicit, and they served their explanatory function by exploiting the power of analogy rather than the rigor of deduction. Duhem confesses himself to be, in this respect, an authentic Frenchman. Commenting on Oliver Lodge's new textbook of electrical theory, he remarks:

In it there are nothing but strings which move around pullies, which roll around drums, which go through pearl beads, which carry weights; and tubes which pump water while others swell and contract; toothed wheels which are geared to one another and engage hooks. We thought we were entering the tranquil and neatly-ordered abode of reason, but we find ourselves in a factory.

Nor (Duhem argues) does this represent merely a temporary fad on the part of these particular English physicists. The habit of organizing physical ideas in terms of concrete analogies, rather than in abstract, mathematical form, is deeply rooted among English scientists, and represents the application within the scientific area of an even broader habit of mind, whose influence ranges over much larger regions of cultural and intellectual life. He compares this contrast between British and French patterns of thought in science with the contrast between Shakespeare and Racine, that between the *Code Napoléon* and the

British tradition of Common Law, and that between the philosophies of Francis Bacon and René Descartes. The *ésprit géometrique* is a part of the French intellectual inheritance, in its widest terms; and this has served to influence the selection-criteria by which French scientists choose between rival hypotheses, just as it has served to influence so many other aspects of French intellectual life. Conversely, the habit of thinking in terms of particulars, and considering them in intuitive and imaginative terms—that *'esprit de finesse* which Pascal contrasted with the *ésprit géometrique*—has been equally characteristic of British habits of thought.[6] As a matter of historical fact, accordingly, the considerations bearing on the "establishment" of novel scientific hypotheses just cannot be stated in a form which will be *absolutely* invariant as between different epochs, different nations, and different cultural contexts. As an aspiration or ideal, such an absolute invariance may be something worth aiming at; but it has never been entirely realized in fact.

Nor is this solely a matter of historical fact. To go further: there are reasons for questioning whether such an ideal, absolute invariance is even attainable. For the processes of "proving," "establishing," "checking out," and/or "attempting to falsify" the novel ideas up for discussion within science at any time are *themselves* subject to a historical development of their own. In a striking series of papers, Dr. Imre Lakatos has demonstrated that our concepts of "proof" and "refutation" have been subject to a slow, but definite and inescapable historical evolution *even within pure mathematics.* What counted as a proof or a refutation for Theaetetus or Euclid, for Wallis or Newton, for Euler or Gauss, for Dedekind or Weierstrass cannot be represented in terms of some unique, eternal, historically unchanging, logical pattern. On the contrary, throughout the history of mathematical thought, the concepts of "proof" and "refutation" have themselves been slowly changing: more slowly (it is true) than the content of mathematics itself, but changing none the less.[7] And if this is true even within pure mathematics—which of all disciplines can most plausibly claim to illustrate the eternal virtues of a formalized logic—must we not suppose that the criteria of "verification," "establishment," and the like in natural science also have undergone a similar historical development?

At this point we can make explicit the intellectual model toward which this discussion has been leading us. For, in the course of expounding all these considerations, we have fallen again and again—quite naturally—into the vocabulary of organic evolution. Science develops (we have said) as the outcome of a double process: at each stage, a pool of competing intellectual variants is in circulation, and in each generation a selection process is going on, by which certain of these variants are accepted and incorporated into the science concerned, to be passed on to the next generation of workers as integral elements of the tradition.

Looked at in these terms, a particular scientific discipline—say, "atomic physics"—needs to be thought of, not as the contents of a textbook bearing any specific date, but rather as a developing subject having a continuing identity through time, and characterized as much by its process of growth as by the content of any one historical cross section. Such a tradition will then display both elements of continuity and elements of variability. Why do we regard the atomic physics of 1960 as part of the "same" subject as the atomic physics of 1910, 1920, . . . or 1950? Fifty years can transform the actual content of a subject beyond recognition; yet there remains a perfectly genuine continuity, both intellectual and institutional. This reflects both the master-pupil relationship, by which the tradition is passed on, and also the genealogical sequence of intellectual problems around which the men in question have focused their work. Moving from one historical cross section to the next, the actual ideas transmitted display neither a complete breach at any point—the idea of absolute "scientific revolutions" involves an oversimplification[8]—nor perfect replication, either. The change from one cross section to the next is an *evolutionary* one in this sense too: that later intellectual cross sections of a tradition reproduce the content of their immediate predecessors, as modified by those particular intellectual novelties which were selected out in the meanwhile—in the light of the professional standards of the science of the time.

An "evolutionary" account of scientific change puts us in a position to re-interpret the spectrum of questions we constructed for ourselves in the preceding section. At one extreme, we saw, the *volume* of new intellectual innovations is highly sensitive to external factors: the relevant questions correspond, in the zoological sphere, to questions about the frequency of mutations within an organic population, and mutation-frequency too is highly sensitive to external influences such as cosmic rays. At the other extreme, the selective factors by which new ideas, or new organic forms are perpetuated for incorporation into the subsequent population, arise very much more from the detailed interaction between the variants and the immediate environment they face. At this level, considerations of an external kind—whether to do with cosmic rays, or with the social context—lose their earlier importance. Now the only question is "Do the new forms meet the detailed demands of the situation significantly better than their predecessors?" And those demands have to do predominantly with the narrower issues on which competitive survival depends.

Does the historical development of a science ever fit this evolutionary pattern perfectly? Can we use it as an instrument for analyzing scientific growth with any confidence? There is no point in making exaggerated claims for the model at this stage. Rather, we should explore its implications in a hypothetical way, to see whether it yields abstractions by which the patterns of scientific history can be more clearly described. Suppose, then, that there *are* certain phases in the history of scientific thought which, for all practical purposes, do exemplify the evolutionary pattern expounded here. Suppose, that is, that there *are* certain periods of scientific development during which all significant changes in the content of a particular science were in fact the outcome of intellectual selections, made according to strictly professional criteria, from among pools of intellectual variants from a previous tradition of ideas. In such a case (to coin a word) we may speak of the scientific tradition in question as a *compact* tradition. Other traditions, which change in a less systematic way, can, by contrast, be referred to as more or less *diffuse*.

Evidently, to the extent that it presupposes our model, the concept of a

"compact tradition" has the status of an intellectual ideal, having the same virtues
and limitations as the concept of an "ideal gas" or "rigid body" or "inertial frame
of reference." We are not obliged to demonstrate that *all* scientific changes whatever conform to this ideal, any more than we need demonstrate that *all* material
bodies are "perfectly rigid," or all actual gases "ideal." Still, if we find as we go
along that the notion of a "compact tradition" can be used to throw light on a
variety of historical processes within the development of science; and if we find
that the deviations from this pattern can, in their own ways, be explained quite
as interestingly and illuminatingly as examples of conformity to it—if "diffuse"
and "compact" changes are equally significant in their own ways—in that case,
we shall be entitled to conclude that the notion is justifying itself. A full discussion of this topic, however, will have to wait for another occasion.[9]

Certainly, we must concede, there are clear instances in which the actual facts of
scientific development do *not* fit our basic pattern at all accurately. Notoriously,
the historical development of some natural sciences has included, e.g., cases in
which the intellectual variants available for discussion at a given time were not
adequately checked or tested, and for many years went—so to speak—"underground": a classic instance of this is Mendel's theory of genetical "factors." In a
sense (one might say) Mendel's theory represented an intellectual variant available within the pool, but one which was overlooked and so failed to establish
itself for more than 35 years. Yet, on second thought, one may inquire:
"On its first presentation, was Mendel's novel theory really introduced into the
general pool of available variants at all?" Was it (that is) put into effective
circulation among professional biologists in such a way that its virtues could be
properly appraised? Arguably, this did not happen: the very limited contact
between the Abbé Mendel and other theoretical biologists in his time shunted his
variant off into a corner, where it could not demonstrate its merits in free
competition with its rivals.[10]

  Again, the development of scientific thought includes occasional phases of a
kind which have no obvious analogy in the sphere of organic evolution. For
instance, a kind of hybridization sometimes takes place between different
branches of science, so giving rise to brand-new specialities, with subsequent
genealogies and histories of their own: the most striking recent example of this
was the emergence of molecular biology around 1950, through the cross-
fertilization of crystallography and biochemistry. By itself, our model of a
compact tradition does not give us the means of analyzing or understanding such
a hybridization.

  Yet again, other fields of intellectual inquiry—known as "sciences" at any rate
to their participants—develop in a way which scarcely exemplifies at all the
orderly, cumulative pattern characteristic of a compact tradition. In sociology,
for instance, the ideas of any one generation seem to have more in common with
the ideas current two generations before than with those of the intervening
generation. There is a kind of *pendulum-swing* in the ideas of the subject, by
which, e.g., "historical-evolutionary" (or "diachronic") patterns of thought alter-
nate with "functional" (or "synchronic") patterns of thought. The latest phases

in the work of Talcott Parsons, for instance, thus recall the ideas of sociologists before 1900, rather than those of sociology during the interwar years.[11]

Once again, however, these criticisms may not represent so much *objections* to our model of a "compact tradition"; rather, they may indicate merely the need for further *refinements* to the model. After all, the very fact that the intellectual tradition of theoretical sociology lacks that compactness which one can find within (say) atomic physics is itself a significant fact. Perhaps there are quite genuine reasons, both intellectual and professional, why sociological theory should not yet have *acquired* the maturity required to guarantee such a compactness and continuity. And perhaps, in his own time, Mendel's ideas inevitably remained "recessive," just because their author was effectively isolated from the rest of professional biology. If that were so, the failure of genetics and sociology to conform to our ideal of a "compact" tradition would do as much to confirm the relevance of that ideal as the actual conformity of more mature and established sciences.

The prime merit of the model expounded here is this: it focuses attention—in a way dispassionate and abstract accounts of the history of "scientific thought" tend not to do—on the questions, "*Who* carries the tradition of scientific thought? *Who* is responsible for the innovations by which this tradition changes? *Who* determines the manner in which the selection is made between these innovations?" And these questions lead one to examine the crucial relationship, within the larger process of scientific change, between the individual scientific innovator and the professional guild by which his ideas are judged. Just how far afield a study of this crucial relationship can lead us is another story, which we cannot go into here; but one point at any rate must be noted.

According to one widely accepted picture of science, the fundamental advances in our knowledge of Nature have all come about through Great Men *changing their minds*—having the honesty and candor to acknowledge the unexpectedness of certain phenomena, and the courage to modify their concepts in the light of these unforeseen observations. This picture of science as progressing through the successive discoveries of Great Men is an agreeable and engaging one, if what Science requires is folk-heroes to populate its Pantheon; yet a little reflection on the actual structure of scientific change may justify one in questioning its accuracy. Indeed, it is a matter for debate how far great scientists *do* in fact ever change their minds; and the actual historical development of Science would—arguably—have been very little different, even if no such "mind-changes" had ever taken place.

Consider, for instance, the work of Isaac Newton himself. We tend to think of Newton as the great intellectual innovator, yet it is worth reminding ourselves how little the basic framework of ideas within which he operated changed between the years of his youth and his old age. The final *Queries,* added to later editions of the *Opticks,* serve substantially to work out in more detail, and provide fresh illustrations of, ideas which had been present in rough, if embryonic form even in his earliest speculations. True, for a while in middle life, having discovered how easily such hypotheses could generate bitter contention with his colleagues, Newton soft-pedaled his thoughts about the ether, and concentrated on less disputatious matters. Still, it is a closer approximation to the truth to represent Newton's intellectual development as comprising the progressive ramification of a fundamentally unchanging natural philosophy, than as involving a series of daring intellectual changes and reappraisals. The great and real change for which Isaac Newton is remembered was that between his own ideas and those which he inherited in his youth from his predecessors. The crucial change, that is to say, was a change *between* the generation of Newton's predecessors and Newton's own generation, rather than a change *within* the intellectual development of Newton himself.[12]

Bibliography

1. D.J. De S. Price, *Little Science, Big Science* (New York: Columbia Univ. Press, 1962).
2. Don K. Price, *Government and Science* (New York: New York Univ. Press, 1954), and *The Scientific Estate* (Cambridge, Mass.: Harvard Univ. Press, 1965).
3. See, for instance, J.D. Bernal, *Science in History* (London: Watts and New York: Hawthorn, 1954) sec. 9.6.
4. *Cf.* Joseph Needham, *Science and Civilisation in China,* Vol. 3 (New York: Cambridge Univ. Press, 1959) esp. sec, 20(c)(2), pp. 186 ff.
5. Michael Polanyi, "The Republic of Science: Its Political and Economic Theory," *Minerva,* Vol. I, No. 1 (Autumn, 1952), pp. 54–73.
6. Pierre Duhem, *The Aim and Structure of Physical Theory* (Eng. tr. P.P. Wiener, Princeton: Princeton Univ. Press, 1954), ch. IV, pp. 55–104.
7. Imre Lakatos, "Proofs and Refutations," *British Journal for Philosophy of Science,* Vol. XIV (1963–64), pp. 1–25, 120–176, 221–264, and 296–342.
8. Even Thomas S. Kuhn, who argued so persuasively for the idea in *The Structure of Scientific Revolutions* (Chicago: Univ. of Chicago Press, 1962), now seems to be retreating from the implications of his own earlier position: see, e.g., his paper "Logic of Discovery or Psychology of Research" in the forthcoming collection *The Philosophy of K.R. Popper* (ed. P.A. Schilipp), The Library of Living Philosophers Series.
9. Discussed in Part I of my forthcoming *New Inquiries into Human Understanding.*
10. E.B. Gasking, "Why Was Mendel's Work Ignored?", *Journal for the History of Ideas,* Vol. XX (1959), pp. 60–84.
11. *Cf.* J.W. Burrow, *Evolution and Society* (New York: Cambridge Univ. Press, 1967).
12. Newton's final account of his natural philosophy, as expounded in Query 31 of the third edition of Newton's *Opticks,* strikingly resembles ideas adumbrated in his earliest notebooks: on this point, see Isaac Newton, *Mathematical Principles of Natural Philosophy* (ed. F. Cajori, Berkeley: Univ. of California Press. (1934), esp. note 55, pp. 671–679.

ROBERT HODES

Aims and Methods of Scientific Research

Robert Hodes was a professor of physiology at the Mount Sinai School of Medicine in New York City. He made significant contributions in at least three areas: in the physiology of the neuromuscular system of man, the physiology of integrative mechanisms of the central nervous system, and the physiology of sleep. Hodes begins this essay by defining science as the ordered knowledge of natural phenomena and emphasizes the *dynamic* character of natural processes and their inevitable interrelations. He criticizes the tendency in much modern science to view phenomena as being isolated and static. Such an approach, Hodes argues, results in a great loss to science.

Dampier-Whetham defines science as an ordered knowledge of natural phenomena or processes, and the interrelations between them. It is worth emphasizing the major points given in this definition: (1) science is ordered knowledge; (2) the knowledge encompassed concerns natural phenomena; and, (3) science deals with interrelations between phenomena or processes, thus indicating a *dynamic* body of knowledge.

The ultimate aim of science is to discover order in nature. All the methods of science are basically attempts to discover such order. For to discover order in any class or group of processes is to explain them, that is, to make them clearer and more intelligible. Conversely, to explain anything is to indicate its place in some orderly system. And in a broad sense it may be said that explanation generally takes the form of tracing the one in the many, or identity amidst differences. The subject matter of science is the real world and in this respect differs from the material covered by Metaphysics.

One of the commonest hazards of scientific study is the failure to recognize the dynamic character of the phenomena being examined. Natural processes are never to be considered in isolation, or as fixed, but rather as related to other phenomena and in a constant state of flux. This is true even in relatively simple situations which can be treated by precise, mathematical formulations. An example is that in which a liquid is in equilibrium with its gas phase in a covered container. Thermodynamics, for its own analysis (and with full knowledge of the limitations of its method of inquiry) treats this system as though it were in equilibrium, that is, stable or at rest. But the fact is that the system is in motion, not at rest. The number of molecules leaving the liquid phase is exactly equal to the number returning to the liquid phase from the gaseous state. Thus equilibrium has been achieved by the interplay of opposing movements in the two phases and is not equilibrium in the sense usually given this term.

What has really occurred is a *dynamic stabilization,* not a *static equilibrium.*

"Aims and Methods of Scientific Research" by Robert Hodes. 1968. Reprinted with permission of The American Institute for Marxist Studies, New York. Pp. 8–18.

And, of course, the level of the dynamic stabilization of the system is related to temperature. One cannot, then, speak even of such a simple concept as vapor pressure without being aware of its dependence on other factors, and without giving proper regard to its establishment by the activity of unresting particles.

How much more difficult, then, it is to conceive of the complex process which is the living organism, without due reflection on the interminable interplay of forces, from within and from without, which go to make up that life. Richet, the French physiologist and student of Claude Bernard—the founder of modern physiology—put the case clearly in 1900: "The living being is stable. It must be so in order not to be destroyed, dissolved, or disintegrated by the colossal forces, often adverse, which surround it. By an apparent contradiction it maintains its stability only if it is excitable and capable of modifying itself according to external stimuli and adjusting its response to the stimulation. In a sense it is stable because it is modifiable—the slight instability is the necessary condition for the true stability of the organism."

In spite of the obvious directness of the above, it seems to me that many scientists are infected with the static approach, to the great loss of science. I need not remind this group that the nervous system plays an important role in the elaboration and course of disease—whether that disease be named classically as schizophrenia, Parkinsonism, or multiple sclerosis, or whether it be classified under the more recently recognized disease states lumped under the name "psychosomatic disorders." Nor do I think I should remind a group of clinicians that the occurrence of a pathogenic organism is not of itself sufficient grounds for the production of disease in any particular individual. But attention in medicine has focused overwhelmingly on the pathogens themselves, rather than on the state of the host who receives them. This, I submit, is basically due to our static approach to the study of disease, our definition of disease in terms of viruses, bacteria, allergic manifestations, etc., rather than as agent-host interrelationships which must vary continually and in infinitely subtle ways, too difficult for us to even begin to comprehend at the present time. Why is it, e.g., that of two individuals exposed to approximately similar quantities and kinds of virus, only one of them develops, let us say, poliomyelitis? Since most physicians and medical research workers appear to take no more than passing notice of what we have expressed as one of the postulates basic to all science (that is, interrelationships) perhaps it would not be irrelevant to suggest that the *nonoccurrence* of disease is of as much, or greater, importance to a rational theory of disease than the identification and intensive study of the infectious agent in the above case.

It appears to me that the time is overripe for a change from research by rule of thumb to research according to scientific principles. And I believe that the major deficiency today is in our lack of theory or concern with theory. Research must be based on sound theoretical foundations, if it is to be consistently profitable; casual, undirected research, instituted without adequate emphasis on basic theory, leads to progress in the field by accident only, if it leads to progress at all. And impressive theory is evolved out of its observational building blocks only when the recognition of the necessity for basic theory is fully grasped. A groping, questioning, incomplete search for the "key" to disease (that is, attempt at rational *theory*) has been published by Chapple, a Philadelphia pediatrician, in the journal, *Pediatrics*. I recommend this paper, not so much for *what* it says, but as a possible orienting discourse on *how* we might, consciously and with singleness of purpose, proceed to free our minds, fettered by the traditions of static, descriptive pathology, and with the work of our hands and brains set ourselves firmly on the tortuous, but rewarding highroad of scientific medical inquiry.

After this digression on the content and aim of science, we may now direct our attention to methods of working with such subject matter. All scientific methods may be divided basically into two major categories: (1) technical, and (2) logical. The technical aspects of science differentiate the various sciences from each other—astronomy from chemistry, and chemistry from social science. For the major difference between the particular sciences lies in the type of measuring instruments or measuring methods used to gather data, or the type of observation employed, or the specific natural phenomena investigated. Competence in dealing with a cloud chamber or a stethoscope, or with particulate matter or a sick patient *alone* distinguishes physicist from physician. Both of them can be serious scientists or both of them can be dilettantes. In this paper we cannot consider the logical approaches involved in scientific methods, since they are all basically the same and fundamental to all scientific methods. Stated simply, the logical components of scientific methods involve reasoning about the facts in order to understand them better and to devise an orderly array between apparently disconnected facts. One thus attempts to place as many different kinds of specific facts into more general terms or laws.

The current popular admiration for science is apparently based largely on the important technical advances that have become the fashion in recent years. Application of new apparatus and concern for new techniques are necessary, of course, to the continuing advancement of science. But as we shall show throughout this paper, the use of new techniques or fascinating equipment without theory and understanding behind the techniques applied to the problems, in and of themselves do not result in fruitful research.

At the heart of the scientific method lies *observation*. Once again, the popular concept of science (and, I might add, the concept of most scientists) is that scientific data result from experiment. As we shall see later, science does not *require* experiment. There are other methods of obtaining valid scientific information. Experiment is based on the fact that one condition or a set of conditions is changed and any effect in the final results is noted and related to the changed conditions. This, of course, is what is known as the control. The basic assumption of the control is that one can abstract or isolate the phenomenon one wishes to study and keep everything else constant. It is most important to recognize that control of an experiment is not easy, although I suspect that most laboratory scientists would tell you that the control is the most common and simplest way of securing real data. For example, in attempting to study the function of the adrenal gland the following experiment has been done countless times: remove the adrenal glands from one animal, take another animal, preferably a litter mate, and do not remove its adrenal glands. Sham operation. You thus use the latter two animals as controls for the experimental one. The animal with the endocrines removed will die whereas the controls will live. But what has the scientist who has performed this experiment actually determined? Nothing more or less than the fact that if one removes the adrenal glands (a procedure) the animal will die (result). Does this mean that the function of the adrenal glands is to maintain life? Or if one removes the cerebral cortex and finds an animal still capable of almost complete motor activity, does this mean that the cortex is unimportant in the control and regulation of movement? Has the experiment with its control solved the problem it set out to solve? No. It has simply related the result of procedure to the final effect—death of the animal or continued motor activity. It has not told us the function of the adrenal gland or the cerebrum anymore than the removal of the gas tank from an automobile, and the subsequent failure of the automobile to run would indicate that the function of the gas tank is to drive the automobile.

But the functions of the adrenal glands will never be discovered by this approach alone. This type of analysis may yield useful information which, when added to other information may eventually provide the answer to the original question. The place of the adrenal gland in the bodily economy will only become apparent when the result of experiment, when clues from disease, when chemical studies and comparative anatomical data, when phylogenetic and ontogenetic observation, when relations between the adrenal and other systems, etc., are collected and reflected upon by the mature scientist with the will and tenacity to forego the easy lure of trivial experiment and face instead with boldness and courage the infinitely more difficult and significant task of attempting to relate in a common system the Babel of voices, all crying "I am the truth." In fact none of them are more than one aspect of truth, and when they proclaim their hegemony over all others, they lose even their claim to being even a part of the truth. For the truth requires that they consider themselves only with reference to the other partial derivative of truth, and not of and by themselves alone.

I point out the difficulty which may attend rigid control because there appears to have grown up in the last period of years what may be called the *cult of the control.* Stringent control is demanded of all experimental work even though there may be no adequate way of controlling the experiment. Some experimenters even require that the controlled experiments be additionally controlled by the application of *statistical* methods. The general thought here is that if a particular procedure is performed and the results are not repeated in every instance or at least in what statisticians have defined as a significant number of times, then the experiments are not considered valid. That is to say, if procedure A (suppose it is adrenalectomy) does not cause death (effect B) in the specified number of times, then A and B are not causally related. The tendency to exaggerate the importance of a statistical pursuit appears to be true in biology as well as in other sciences. Tables are kept of the number of times phenomenon A is associated with phenomenon B and statistical maneuvers performed on the "raw data," as for example the number of patients who improved after some drug therapy and the number who did not improve.

I might point out that the statistical approach, far from being a fundamental one, is rather a makeshift. It should be utilized only when there exists no better way of handling the data. And I suspect that frequently there is no better way of handling the data because we have not searched diligently enough for other directions. For once the fundamental relationship between process A and process B is established, statistics become unnecessary. Then the occurrence of B with A can be predicted accurately without the need for tables or standard errors of the differences between two means. The ancient Egyptians kept records of solar and lunar eclipses and these tables at the time were useful; for they enabled these people to note certain cycles on which they could base anticipations of eclipses, although they did not understand the reasons for them. But now, since laws of the occurrence of the eclipses have become known, there is no need for statistical records. They can be foretold with accuracy and confidence. In fact, you read in your local newspaper the exact time to the minute, when the next eclipse will start and when it will end.

In a similar fashion there is no need to keep ponderous statistics on the efficacy of insulin in diabetes since the relationship between this disease and a deficiency of the pancreatic secretion makes such correlations unnecessary. Indeed there is danger of overemphasis on statistics to such an extent that false conclusions may be drawn or valuable relationships missed for the too rigid application of this approach. An example of this occurred in our laboratory recently when a colleague and I were studying the results of stimulation of sub-cortical structures on the electrical activity of the cerebral cortex. A particular type of electrical response, which I will not go into here for lack of time, occurred two times and did not occur twelve other times when fourteen successive subcortical stimulations were presented. The stimuli were apparently identical, the animal was the same, and no more than 20 minutes elapsed between the first and last stimulus of the sequence. Statistics would have indicated that the effect produced two times was not significant and had occurred by chance— which is another way of saying that one did not understand the causal relationship between the stimulus and the response. Closer examination, however, revealed why the effect had happened twice and not the other twelve times. The occurrence of the effect on two occasions could be related to the state of the cortex. This was considerably different from those times when the stimulus did not produce the given effect. An improper concept of the function of statistics and their rigid application might tend to emphasize the static features of the experiment—the end result of stimulation in this case—and tend to minimize the importance of examining those dynamic relationships which in the long run may be more important than the "poor" statistics.

I mention certain difficulties with the experimental techniques not to belittle the great contributions which experiment has yielded to science. I present it only to indicate that experimental evidence is only one part of the total method of science and must always remain so. I mention it also because there is a tendency, particularly amongst laboratory men, not to accept anything but experimental evidence. Walshe, an eminent British neurologist, deals with some of the problems I am trying to discuss here today, and from which carefully reasoned and impeccably presented work I have drawn extensively in this paper. In deploring the attempts of physiologists to explain function on the basis of simple physiological experiments alone, without recourse to other available data, frequently controverting the experimental results, he ascribes the insolence of the laboratory man to "his unwillingness to accept as proved anything that is not experimentally proved: that is his equation of the large term 'proof' with the smaller term 'experiment proof.' " I mention it, finally, because I am sure that the deliberations at the subsequent sessions of these meetings will show that possible research is being done and its scope should be expanded in psychiatry as in other fields, even though experiment may not be the main source for accumulating data in this branch of science.

So overwhelming has been the emphasis on experiment that the outstanding neurologist of his day, Hughlings Jackson, felt constrained to argue against such intemperate views. He said: "A collection of numberless facts, however accurately gathered, is not a real experience. Unless a man can put a particular phenomenon he himself sees under more general laws, or unless he tried to do so, he scarcely can be said to know or to be studying a thing in any very valuable sense." It is pertinent to reflect on these words of Jackson written almost one hundred years ago in considering the appalling plethora of disconnected and obviously unconsidered experiments that fill our scientific journals. Equally discouraging is a perusal of doctoral theses of potential young scientists who have apparently performed their experiments "because no one ever did them before."

There seems to be a growing tendency to use more and more powerful microscopes, to use bigger and better electronic devices without giving much thought to the reason why one is using them or what one expects to find from the use of these important new instruments. The question may reasonably be asked: are these productions—aimless, thoughtless and trivial—are these experiments science? In my opinion, the answer is a categorical no. For if we refer to our original definition, there has been no attempt in these experiments to relate phenomena, no attempt to place the phenomenon under observation into more general schemes or in ordered relationship with other phenomena.

We have thus far arrived at the point where we must accept the fact that experiments, employing complicated apparatus, ever more accurate and impressive as instrumental "tours de force" however adequately controlled or statistically significant, do not constitute *per se* science or the scientific method. Experiment yields information, but experiment alone does not produce science. Nor is experiment the only means of obtaining scientific facts. Careful observation of natural phenomena is another way of securing serviceable facts of science. We need mention nothing more to prove this than to remind you that one of the truly great scientific theories of all time was elaborated without recourse to a single experiment. I refer here to Darwin's theory of evolution. Darwin observed nature, recorded his observations with meticulous care and astounding prescience and thereby gained the raw material for his theory. But he was unable to intervene in or alter the course of the process he was examining, and hence he did not experiment. One might also mention that the sciences of geology and astronomy in essence are observational sciences without benefit of experiment.

It is high time that physiologists and physicians alike recognize the fact that experimental physiology does not have a monopoly in its attempts to uncover the functions of the nervous system. As Walshe has pointed out, indeed, many of the tenets of experimental neurophysiology do not agree with observation made in the clinic. And it is fatuous to believe that the experimental approach is the only correct one or even always the more important one.

We have stated that observation lies at the very heart of the scientific method for from observation come the primary data of science. The next step in the logical construct of science demands that the scientist now apply various types of reasoning to the data he has obtained, either by experiment or by other means. If a relation can be found between the new material he has uncovered and already existing material he will attempt to include his work into the general framework. Every new set of facts which essentially agrees with current theories, strengthens the existing theories. However, if the scientist is faced with data of his own which have no apparent or immediate relation to similar pre-existing data he must attempt to find the key factors which relate his own work to that material already attained. If he fails to find any such correspondence between his own evidence and that required by existing theory he must either look to the adequacy of his own research or the validity of previous material, or even be prepared to abandon earlier theory. By combining different types of information he may produce a hypothesis in which the particular data obtained are incorporated into a more general scheme. Such a hypothesis will usually lead to further experiment or observation and the verification or the destruction of the primary postulates. Hypothesis is thus at the base of prediction, one of the fundamental attributes of all scientific methods. A correct hypothesis will allow a correct prediction. When theory no longer adequately conforms with observed phenomena it must either be revised or abandoned.

In current scientific writings a sharp line is frequently drawn between "pure" and "applied" science. Biology is considered pure, medicine but one application of pure biological science. This distinction is a meaningless and indeed a very harmful one. For, as Walshe has pointed out in the article referred to above, it is quite clear that the function of the brain in the control of motor activity may well have received more illumination by the use of the "applied" science (medicine) than by the use of the so-called "pure" science of neurophysiology. This artificial distinction between "pure" and "applied" science perhaps has arisen from the desperate realization that science, particularly those branches commonly called "applied," are without rational theory. Conventional imagery produces the be-spectacled, unworldly pure scientist, racking his mammoth brain over weighty matters of theory. He is as unconcerned over the results of his labors as he is about his material well being. He is immersed in science for its own sake. He foregoes the pleasures and comforts of life because knowledge alone is his sustenance. He seeks truth for its own sake and for his own satisfaction, and not for its intrinsic value to mankind.

But the applied scientist is conceived of as a good-natured fellow, not a very serious thinker, always the practical man, always ready to put some theory to use, always willing and able to transform ideas into dollars. He thus has out-

distanced his alchemist predecessor, who at least had to purchase the base
materials which were to be converted to gold. Although there may be some
modicum of truth in this contrast in the actuality of present-day science, this is
not the fault of science, but of its devotees. It need not be so; it should not be
so. For such a distorted view denies the essence of science—that hypotheses,
generalizations, or laws of nature must stand in conformity with actuality,
or else be declared false. The theorist who scorns practical consideration is as
incomplete in his theory as the pragmatist who denies the value of theory is
lacking in his practicality.

   Theory suggests practice, which in turn suggests further or better theory, and
vice versa. Theory and practice not only supplement each other—they are integral
and essential parts of truth. The identity of the theoretical and the practical
was recognized by Francis Bacon, the first important figure in the philosophy of
modern science. An ardent exponent of the value of science in the fight to upset
Aristotelian dogma so tenaciously upheld by the clergy of feudal England,
Bacon's writings were a major factor in inspiring man to release the productive
forces which ushered in the present era of technical and scientific advance. Bacon
states: "The improvement of man's lot and the improvement of man's mind are
one and the same thing." And conversely, "To be ignorant of causes is to be
frustrated in action." And again, science "should be designed not for mental
satisfaction but for the production of work."

   A more recent statement of this position, directly applicable to biology and
medicine is the following eloquent plea of Wilfred Trotter: "When ideas are
freely current they keep science fresh and living and are in no danger of ceasing
to be nimble and trusty servants of truth. We may perhaps allow ourselves to say
that the body of science gets from the steady work of experiment and observation its proteins, its carbohydrates, and sometimes—too profusely—its facts,
but that without its due modicum of the vitamin of ideas, the whole organism
is apt to become stunted and deformed, and above all to lose its resistance to
the infection of orthodoxy. To the experimenter immersed in his research and to
the clinician struggling with the load of experience and the needs of his patients
it may seem impractical to concern ourselves with the theory of medical knowledge. On the other hand, it is perhaps the lack of rational doctrine and of a
general interest in the problems of method that has made medicine the scene of
so much disunited and contradictory effort, and helped to put it down from its
historical position as the mother and nurse of science."

Recommended Readings

Boas, M., *The Scientific Renaissance* (New York: Harper and Row, 1962).

Broglie, L. De, *The Revolution in Physics* (London: Routledge and Kegan Paul, 1953).

Burtt, A.E., *The Metaphysical Foundations of Modern Science* (New York: Doubleday, 1932).

Butterfield, H., *The Origins of Modern Science* (New York: Macmillan, 1961).

Hall, A.R., *The Scientific Revolution* (Boston: Beacon Press, 1954).

Heisenberg, W., *Physics and Philosophy* (New York: Harper and Row, 1959).

Lakatos, I., ed., *Criticism and the Growth of Knowledge* (New York: Cambridge University Press, 1970).

Needham, J., *Science and Civilization in China,* 5 vols. (New York:Cambridge University Press, 1954- —).

Randall, J.H., Jr., *The Making of the Modern Mind* (Boston: Houghton Mifflin, 1940).

Sarton, G., *Introduction to the History of Science,* 3 vols. (Baltimore: Williams & Wilkins, 1927-1948).

Science and Technology:
The Moral and Political Issues

Science and Technology:
The Moral and Political Issues

Introduction

The scientific discoveries that have occurred in this century and the technological applications of them are of a magnitude unparalleled in human history. The effects of these developments on man are also unprecedented. As we pointed out earlier, science and technology served in many instances to liberate man from constraints and ravages of the past. Advances in medicine have eliminated diseases and extended man's life span. Developments in industrialized nations have provided new modes of production of goods, food, and services, and new means of transport and communication. These innovations and many others are so familiar to us that it is tempting to view scientific and technological progress as being entirely beneficial.

Yet this is not true. One need only reflect on the parallel developments in weapons and other instruments of war to realize that man's condition on this planet is far less safe than it was a century ago. The threat to our environment, brought about by technology and industrialization, would have been unimaginable to nineteenth century man. Technology has brought new and powerful methods for the domination of man by man, and it can be argued that in many ways man is less free today than he was in the past.

And what of the future? Advances in science and technology hold the promise of utopia if put to wise use; toil, drudgery, and disease would be eliminated. Man's condition would be ameliorated, liberated, and made peaceful. On the other hand, used improperly, science and technology offer a dismal prospect, a "Brave New World," or a "1984," or worse; and with it, the possibility of the destruction of the human species as we know it, through nuclear war or through misapplied eugenics and genetic engineering.

Clearly, then, the *ethical* and *political* issues raised in and by science assume greater importance than ever before. We can no longer ignore the prospect of subjecting the applied sciences (and possibly the pure sciences, as well) to some form of social and ethical constraint. Modern science is, after all, a cultural phenomenon: it cannot remain indifferent to the uses to which it will be put. If the exponents of science ignore the broader ethical problems of our culture, they may inadvertently contribute to their own destruction—and to ours. Science and scientists can no longer be indifferent to the problems of human values.

But, what values shall provide the needed guidance for future technological development and use? Who shall make these judgments; and how? To what degree might we be forced to reconsider our own values—personal and cultural? And finally, what changes must be wrought in the basic organization of society to provide for the possibility of a liberated and liberating science?

These are some of the questions discussed in this section.

ROBERT S. COHEN

Ethics and Science

Robert S. Cohen is Chairman of the Department of Physics at Boston University and is co-editor of *Boston Studies in the Philosophy of Science*. He has written and lectured widely on science and culture. Are science and ethics related, and if so, what contributions can science make to ethical thinking? Can scientific research enable us to evaluate moral theories? Do the ethical codes employed by scientists in the practice of science itself, offer models for ethical codes of mankind generally? These are some of the questions that Professor Cohen raises in this provocative essay.

Speaking about science and humanistic education, which is to speak about the social relations of science, it is useful to address ourselves to the purest of social questions: not only such practical questions as arise in technological change and economic development, in allocation of our finite resources, and in mastering the runaway accumulation of knowledge and of people, but also the most ancient and simplest of questions, how shall a man live with his fellows and with himself? In every civilization we know, Asian, African, European, there have been those men who have said with Socrates that the unexamined life is not worth living. Indeed, the history of those who have been called "wise" is a history of such explicit examinations of life.

How men shall live with one another is the subject matter of ethics. And here I mean to include social ethics, as another name for politics, along with personal ethics. From the beginning of Western thought, certainly in Greece, and perhaps in Egypt and the Mesopotamian Valley, the habits and customs and myths, which constitute the morality of daily life, have been subjected to a professional scrutiny and criticism. In this specialized part of the division of intellectual labor, the priests and the philosophers, the poets and playwrights, tried to establish grounds for reinforcing, or overthrowing, or plainly establishing beliefs about what is good and what is right. The very fact of inquiry into right and wrong was a tremendous advance, for such thinking recognized that there were alternatives from which you might choose, or at any rate your teachers and priests might choose. Freed from the narrow moral vision of a single clan or tribe, men began to think as widely as they travelled. It is likely that this widening of vision was a by-product of international trade, and especially of those remarkable Ionian Greeks of 600 B.C. who sold their goods throughout the Near Eastern lands of Persia and Egypt, who learned techniques and medicines and religions, and then compared and contrasted them for reasonableness, for evidence, for moral satisfaction, and for happiness.

This lecture was prepared for a conference on "Science and Technology and Their Impact on Modern Society," Herceg Novi, Yugoslavia, September 14–23, 1964; and revised for delivery at Merrimack College, North Andover, Massachusetts, November 4, 1972.

The moral vision in European thought had another source, and apparently quite separate, in the Hebrew tribes whose conception of their Gods changed over a relatively short number of generations. From the usual agricultural and nomadic idolatries, and the pantheons of tribal gods, they turned to the idea of One God. More importantly, they went further to conceive two even more striking changes. First, the Hebrews' One God was thought to be God of all the world, and not because this God would lead them to conquest over all men, and Himself to conquest over all other tribal gods. Rather, He is the God of all by virtue of the inherently reasonable notion that the universe as a whole deserved an explanation which the separate gods could only partially offer and then only with incoherence or with unexplained contingent conflicts. The chaos of gods now became a unified (still personified) system; and the system was to be rational. So, for the Hebrew mind a rational center of responsibility for the whole world was established: here is proto-science and early religion.

The second striking Hebrew characteristic was the nonpictorial nature of the new One God. Even at the beginning, with the story of Moses and the burning bush, God was to be a god of ethical qualities, not so much an anthropomorphic material being *with* behavioral qualities of justice and goodness and anger, as He was the being of the ethical qualities themselves, joined together in a mysterious personality to be sure, since only a person could be understood as 'responsible' for the World.

In both these sources, Greek and Hebrew, there is therefore a tendency toward reason combined with morality, of justice dispensed in the light of evidence, of a searching for explanation not by acts of arbitrary powers but either by reasoned godly actions or within patterns of natural sequences. In Greece particularly, then, thought about the actions of men and nature and gods turned self-critical, toward the preliminary questions: how to think, how to be sure, how to recognize truths, or, briefly, how (and how much) can we know? Here in this question of how to establish knowledge, in this long search for no magic key but for a common method to truth, is the first link of science with ethics.

The search was difficult, and the philosophers of many outlooks. There have been idealists like Plato who offered theories of absolute insightful and *a priori* ethics tied closely to their brilliantly fruitful insightful mathematical approach to absolute knowledge of Nature; there were Aristotelian materialists who were successful taxonomic and developmental empiricists as much in their social and political ethics as in their biologically modelled approach to understanding Nature; there were the humanist empiricists like Hippocrates who offered experimental evidence and other lessons from experience for their reasonable scientific estimates of the probabilities or the impossibilities of diagnosis and cure, along with humane and rational advice on the conduct of life in this uncertain, only partly known, but wholly natural world; and among many others, there were the skeptics like Protagoras and Diogenes and Sextus Empiricus who logically criticized the sufficiency of the evidence for every belief about Nature, indeed who demonstrated the untrustworthiness of our sense-perceptions and our rationality alike, and who likewise were cynical about any trans-human moral standards whatsoever for human conduct. Theories of knowledge and theories of ethics were not only analogous; they were closely linked.

Later analogues to these tendencies are evident enough. We have had our Platonists in science seeking logical and mathematical assurances and with these, the ethical certainties of almost mathematically absolute rigor, of Descartes, and Leibniz; and later the Kantian parallel *a priori* of natural and moral laws. We have had the pure empiricists, the experimenters and observers of science, and with them Baconian ethics; and, later, equally empirical observations of moral differences throughout the world with the simple conclusion of ethical relativity. How appropriate for empiricism. And we have those utter skeptics about a reasonable basis for human knowledge, Montaigne and Hume, with their equally skeptical abandonment of any reasonable basis for ethics too. Science and ethics have seemed to march together, throughout the history of civilization.

Traditionally then, philosophy linked knowledge of nature with knowledge of morality through epistemology. And since epistemology arose and evolved primarily by its reciprocal dependence upon science, so ethics was dominated by science. This dominant influence took several forms: conceptions of how to think at all; philosophical conceptions of what constitutes an explanation; practical scientific ideas of what is possible and what is impossible in this world; scientifically known (or allegedly known) facts of human nature, social nature and the world at large; often, too, a scientific restatement of theology so that at several crucial stages the laws of nature were thought to be divine legislation just as the laws of politics and personal ethics. Much of the historical interaction of science and ethics reflects just this ambiguity in the concept of law, in the slow development of the distinction between the laws of men (which may be violated and which then may demand just and presumably equal punishment) and the laws of Nature, which cannot be violated. The idea of causality, so central in the scientific world-view and however subtle it has become recently, came from the ethical notion of retribution,[1] of a punishment fitting a crime and hence of effects equaling their causes with respect to some essential quality.

Yet we must also admit that there have traditionally been major gaps, even conflicts, between science and ethics. For many European thinkers, this was the great gap between nature and the supernatural, between body and spirit, or—more subtle—perhaps between responsible private men and irresponsible and unreasoning public society.[2] Even within the nonreligious outlook of naturalists or atheists, the ancient observation of a gap between fact and ideal was disturbing. It was simple enough to see that knowledge is not necessarily liberating; indeed, in one formulation, what may be used for good by one hand may be used for evil by another. Science deals with facts, we often say, and ethics deals with values, and the gap becomes unbridgeable. Most succinctly, it is said that an *ought* cannot be derived from an *is*, no imperative from a declarative. Shall we formulate this as a contrast of knowledges: knowledge of facts versus knowledge of values?

The matter is clear, even within the elementary ethics of common sense. If we know what we want, then knowledge of facts will help us, either to achieve our goal or to tell us that it cannot be achieved. So science, if conceived as knowledge of the relevant facts, will inform us about the *means* to our ends, but it does not thereby shed light on the wisdom of those ends. And the grim fact remains today that men not only may differ about their ends, from man to man, from class to class, from nation to nation, but also that their ends frequently have been incompatible. It is better, I believe, not to claim too much for science at this time. Not many centuries ago, science was thought to be *inherently* good. Phrasings vary: the truth shall make you free; the unity of the True, the Good, and the Beautiful; if you know the good, you will do it. Indeed, the revolutions of politics and science of the sixteenth and seventeenth centuries incorporated into their various ideologies the maxims of both Socrates and Bacon; knowledge is virtue, knowledge is power. Science is knowledge, and science is good.

But the full truth is bitter. Science is no longer the wholly enlightening ally of human progress that it once seemed to be: and humane men will look warily at any model of a scientifically 'rationalized' social order, with too strict a devotion to facts, with concentrated focus of intellectual resources upon these technical fields which have made possible the mechanization of human life and culture. Whether by a Marxist critique of human relations in present advanced industrial society, or by an existentialist critique of the private individual's isolated situation in modern mass rootless societies, or by religious despair at the prevailing absence of love and of genuine comradeship: by any of these we come to realize again that science alone and by itself is morally neutral and painfully so. It is not automatically, is not spontaneously, a force for good; and hence neither individual nor society can depend upon such a neutral social institution as science. Being itself uncommitted, it is dangerous as well as powerful, irresponsible. Without responsibility, there is no ethics.

Furthermore, the extension of rational science beyond mechanics to the study of society and humanity, to history, is no guarantee of a rational humane commitment within the scientific community, no guarantee of moral wisdom within scientific knowledge. If nuclear bombs were the twentieth century's greatest destructive advance in man's masterly manipulative knowledge of nature, then the extremely rapid achievement of unsurpassed barbarism in Germany in the early 1930's, that is in one of the most civilized nations of Europe, is an achievement of man's masterly manipulative knowledge of human nature. And yet, so many humane thinkers in so many different traditions have known what is wrong. From Jeremiah to Jesus, from Erasmus and Comenius to Diderot and Jefferson and Kant, from Saint-Simon and Marx and John Stuart Mill to Veblen and Dewey and, I believe, from Lenin and Wilson and Gandhi to Franklin Roosevelt and John XXIII and Bertrand Russell, Brecht, Camus and Sartre, all have seen that the problem of the good society is to know how to treat each man and woman as an end, not only as a tool or as an instrument; not as an object alone but as a subject too. The grave difficulty thus far in the history of mankind has been that the material basis for life remains limited and finite, with attendant and apparently unavoidable exploitation of many men and women by a powerful and fortunate few. The ethic of have and have-not peoples remains, the gap continues and astonishingly it widens with technical advances, steadily. But then perhaps more science will help at last?

Will nuclear energy set us free? Free from poverty and hence free from greed and jealousy and even from war and exploitation of men? Will widespread science make us free from idolatry and other illusions, free for truth and curiosity, play and art and love? Indeed, will more and better science, a more completed science of nature, end the ghastly monstrosities which have plagued European civilization with its technology and its science?

These, and connected questions, have stimulated a rejuvenation of thought about ethics in recent decades, by philosophers in many countries and with different viewpoints. And in their several ways they have repeatedly sought to discover how science and ethics are related.

The role of science in ethics is complex. First, scientific *methods* may provide methods of ethical thinking and moral discovery. Second, scientific *investigations* of societies may provide what Spencer called 'the data of ethics.' Third, scientifically plausible *theories* may explain by comparative analysis the historically situated differing moral systems by using psychology and history in the broadest senses. Fourth, scientific *achievements* may determine the scope and limits of responsible moral choice and decision. Fifth, since science is a human enterprise, the *life of science* may itself provide a moral lesson of how men may live with one another, that is, the study of the history of science may show us a scientific ethic. Sixth, science brings us to new choices, new problems, and new circumstances for old problems. These modes of relevance between science and ethics may be labelled as: first, analytic; second, descriptive; third, comparative and causal; fourth, instrumental; fifth, behavioral; and sixth, problematic. Let us examine these briefly in turn.

But you may ask, what is the first of these, 'analytic,' intended to mean? Surely not what is meant by 'mathematical analysis' nor again what is meant by 'experimental analysis' in a laboratory or in the field. It means 'philosophical analysis'; and this phrase presupposes, first, that philosophy is possible at all (which tough-minded scientists often deny, at least until, in their elder years, they retire to write books of philosophy), and further, it presupposes that philosophy has a distinctive method which differentiates it from science. The claim has been made that science supplants philosophy, and the historical record does affirm that the description of the world of nature, and of human society, is no longer to be the work of contemplative philosophers but rather that of patient, observing, and *active* scientific investigators. Physics and astronomy came from natural philosophy; psychology from the philosophy of mind; history (which I believe is a science in its own way) from the philosophy of culture; and sociology, scientific anthropology, economics and the rest of the social sciences have almost entirely supplanted the old speculative assertions of metaphysics.

What is left? What is the function of philosophy aside from the study of its own history, taken as an instructive and at times amusing recounting of prescientific illusions and prejudices? But such irony is doubly wrong. For one thing, our modern ways of scientific knowing should be scrutinized for their own metaphysical presuppositions, gaps, illusions, and prejudices, and in this their linkage with the past will be certain. And for another, what in the past was speculation about metaphysical abstractions often turns out to have been sharply reasoned analysis of human situations.

We have a profound, if simple, example in the history of that fundamental term of European metaphysics: Reason. Ultimately, all the philosophical idealists sought with *a priori* argument, to establish a belief in a rational world-order; and the materialists, both those who followed a dialectic and those who worked in the empiricist or pragmatic traditions, found the social content of metaphysical 'Reason' by means of an elementary transformation which has revolutionary consequences: namely, if we first see that a demand for genuine individual happiness is the content of the metaphysical demand for a rational world-order, then we may comprehend that "the realization of reason is not a fact but a task."[3] The task is to materialize values.

If to make the world reasonable is our task, then, the philosopher must turn from philosophy to the material and causal sciences of physical and human nature on the one hand, and to critical social theory and practical concrete activity on the other. Not idealist utopian analysis, but concrete knowledge is needed for this task; not pure thought but scientific practice and thoughtful criticism. And so, as with Reason, the other metaphysical categories and speculations of the prescientific philosophers and poets, while set forth in strange, abstract, and even irritating languages, deserve respectful study and then reconstruction and reinterpretation.

After such sociological reconstruction of metaphysics (into humane social sciences), what *then* remains to philosophy? Perhaps only logic, but not mathematical logic, which has passed to the mathematicians; it would be a critical logic which remains, and indeed philosophy has only this critical role to play. Never really creative or discovering, philosophy is, as Hegel gloomily reminded himself, the wise owl who flies only at night, when the bright creative life of day has passed. And thus conceived, philosophy is ancillary both to science and to social realities, a Socratic gadfly still, a conceptual irritant, seeking to clarify meanings and techniques, to pose overlooked questions, to relate hitherto unrelated fields, to transcend egocentric and sociocentric predicaments. Max Horkheimer was right, at the depth of Nazi savagery: More than logic, philosophy still has its old tasks, to explicate human knowledge and to criticize the human situation. And, at least in the recent English-speaking countries, the technique is called 'philosophical analysis.'

One of the philosophical analysts, John Wisdom, put the matter this way: "Philosophical progress does not consist in acquiring knowledge of new facts but in acquiring new knowledge of facts—a passage via inspection from poor insight to good insight."[4]

The rigorous analytic techniques of scientific philosophizing have been used to clarify the major issues and terms of ethical discussion too. A slogan for analysis, and its ally logical empiricism, has been one of 'rational reconstruction' of scientific theories and ethical doctrines alike. In ethics we must distinguish means from ends, and know when achieved ends become means to higher ends-in-view; we distinguish purposes, wishes, exhortations, commands, which have easily been confused. And above all, with Charles Stevenson, the meaning of ethical terms has been seen to involve two distinct components, that which implicitly asserts matters-of-fact (whether true, false, or probable in some degree) and that which expresses an emotional significance.[5] There is an 'emotive meaning' as well as a cognitive meaning. In the specialized division of philosophical labor, such analysis of ethics may hope to provide an answer to the puzzling and important question, "What sort of reasons can be given for normative conclusions, i.e., for conclusions about how men ought to behave under specific circumstances?"

Such analysis is scientific in spirit but *limited* to logical and linguistic techniques and to neutral abstractions. With *descriptive* ethics the scientific scope of ethical investigations widens greatly. The history of practical moralities and of ethical theories, and the investigations by ethnologists and sociologists, have by now provided an astonishing range to the phenomenon of moral consciousness. Not merely the breadth but also the subtlety of these moral phenomena provide scope to artists too, for beyond the questionnaires and surveys of spontaneous awareness, beyond social science, we can also recognize moral situations, moral responses, and moral responsibilities through novels and drama. Penetrating description of the distinctively human situation is the scientific function of literature, for it is a cognitive function: to reveal how the world feels to this man in this place. Through a union of subject and object in art, this feeling of a real world by a real subject (however fantastic the fiction may be, its truth remains genuine), we learn of ethical phenomena scientifically, through literature as through anthropology. As a final descriptive factor the existentialist and phenomenological philosophers have stressed the psychological phenomena of individual responsibility as against sociological or historical accounts of moral behavior.

A further stage which uses all the resources of social and psychological science is *comparative*. Examine the moralities of different times and places, different peoples, different generations within one people, differing economic and racial and linguistic circumstances within one society; explore the historical development and cross-cultural diffusion of moral ideals; and then certain conclusions may be drawn by such comparisons, if drawn scientifically rather than romantically and impressionistically. Thus, in the most direct though elementary way, we might be freed from ethnocentrism, that constricting vision which sees our Western ethical rules as a universal standard. We will be assisted in a more valid search for ethical invariants throughout the variety of conditions of life on earth without any *a priori* commitment to a specific set, or even to the existence of such. Further, by setting forth the "different ways in which human beings have tried to do similar jobs" we can be aided in a practical evolution of criteria for success. In MacBeath's phrase, we examine "experiments in living." Comparative studies offer the soundest basis for understanding the relation between ethical behavior and historical culture, and for appreciating the extraordinary inventive and creative possibilities for stable societies, indeed for appreciating the fine title of Gordon Childe's history of early societies, *Man Makes Himself*. We see that what might be legitimated as goals will be not only deeply rooted in the immediacy of daily life and recent events but also may transcend any local situation; a scientific insight which describes what may be called, with Paul Meadows, a situational dialectic.[6]

It is easy to see social and historical content in ethical theories, for they usually state and guide the interests of a class or church or other group. In this sense, comparative analysis reveals that ethical theories themselves are not just about values; they *are* values, since they *consist* of the goals and interests of certain groups.

Does a given ethical theory rationally justify a pattern of behavior? Is it true? Whether true or false, any operative pattern simply is the values of its adherents; and it provokes us to try causal-historical analysis, to a sociology of ethical ideas. Indeed the social functioning of ideas and the social determining of ideas are the crisscross intersection of science-as-truth and ethical values-as-interests, both investigated by scientific comparative studies.

Probably Hume and Marx in Europe and Veblen in America were the clearest voices to call for prompt social-scientific investigation into the causes of moral sentiments. What were the origins, the variables of function and displacement, the effects and complexities of truths and distortions in human goals? Remarkably, both Marx and Veblen combined a passion for empirical observation and hypothetical theory, that is, a scientific base for criticism, with an equally passionate moral critique of their own times. And when Marx said that the criticism of religion was the foundation of all criticism, he was calling direct attention to the hypothesis, as Reinhold Niebuhr phrased it, that every "claim of the absolute is used as a screen for particular competitive historical interests."

Now this hypothesis can only be established by empirical investigation. Indeed, there is an instructive example of the interaction of causal social science with ethics in the Marxist critique of religion; for, as the French priest Henri Chambre remarked aptly, Marx was not battling gods but idols and fetishes: "Marxist atheism does not believe in God but in men . . . it disputes the supremacy of things (idols) over men."[7] Hence we might say, the central *scientific* problem of the theory of knowledge, a problem whose solution is the basis of liberated thinking, is the nature of idol-producing thought, whether it is thought about nature or about values. And so once again, the relevance of science to values is established.

Just as scientists may study ethical codes, so they may study the specific ethical code of *scientists*. Let us idealize but not the biographies of our teachers, friends and colleagues, nor even of great thinkers and experimenters. If we were to do that, I suppose the evidence would only show the same moral behavior that other men and women have had: superstitions, opportunism, wishful thinking, bandwagon fashionable enthusiasm, prejudice against the new or the foreign, all these and more can be found by studying the behavior of scientists with a slightly jaundiced eye; and of course we probably have our share of merit also. But what I mean to do is to recall the ideal which scientists do know about. The life of science, the ways we live with each other when we are true to the unimpeded knowledge-seeking goal of science, is characterized by an ethic with notable positive features. We form a democracy whose citizens decide what shall be the policy, what accepted as truth for guiding the commonwealth. The citizens who are the voters have an educational test, a special literacy test, perhaps the Ph.D. degree or another form of introduction to participation in the literature and discussion within the forum of ideas and techniques. In this situation, which is political despite its lack of formal structure or national boundaries, we have leaders but we choose them ourselves. And each of the leaders is ever open to challenge and then to replacement. Shall I say the leading ideas are replaced rather than the leaders?

But the matter ought to be put a bit more strongly. Science does have a unique fusion of obligation and rebellion. We cannot have science without realizing our obligation to past workers and current colleagues; and we know there will be no science if we do not honor our obligation to future scientists; we may choose not to have science, but insofar as we are scientists, we act with devotion to social cooperation. And we also seek social criticism since the very obligations to others which ensure the existence of science also ensure that mutual criticism occurs. Indeed the cumulative nature of science includes a recurring theme of revolt, since at the crisis points of investigation and interpretation, we do honor to our masters, teachers and heroes precisely by coming to reject them, learning to repudiate them. In its inner character, science is both objective and tentative, a temporary knowledge about persisting realities. And the political scheme is simple enough. It is a democracy, almost naively so in spirit, a democracy of the sort described by Thomas Jefferson in the early years of the United States. We scientists do not have formal elections, much less regularly scheduled ones, but we do have that plausibility of a true democratic practice: we give an idea, or a theory, or a technique, a test; we choose some men and their proposals and let them run the affairs which are on our agenda and after a while, we test them against our experience, and decide whether they are right or wrong, wise or foolish, the best likely or the least likely to succeed. And usually we replace them. And while Einstein replacing Newton does so with the greatest respect for Newton, there is yet psychological interest in the manner by which science utilizes the revolt of young generations against the old, for positive ends.

This scientific democracy has an additional quality which should demand respect: social collaboration is combined with extraordinary respect for individual work. If ever the conflicting claims of classic bourgeois individualism and classic socialism will be reconciled in a fully healthy society, it will reflect this beautiful legitimacy of *independence and interdependence* within science.

Along with this synthesis there are several other contrasts. In science, we combine subjective attitudes with objective demands, for example, esthetic delight with a demanding reasonableness. We combine beauty with utility, pride with modesty. We combine authority and leadership with private judgment and constant individual criticism. And we should treat each other with respect. Despite violation by pride and other weaknesses, the ethic of the international community of scientists is persistently known. *And the ethic of science is the democratic ethic of a cooperative republic.* Insofar as its own character may be a model for other human enterprises science has therefore a noble relation to ethical behavior. Indeed, if science teachers would bring to their students a conscious attention to these factors in the history and current practices of science, those students would be morally educated as well as technically.

Nevertheless praise of science must be limited. This internal ethic has not always characterized the external social relations of science. In fact, science is in society as a subordinate instrument of power. If we are to understand the place of science in society, we must turn to the history of the several social functions of science. Here milieu and context are decisive. Moreover, just as we have seen how glorious or terrifying has been, and yet may be, the impact of science on society, so also we must assess the fact that society has its decisive impact upon science. It provides resources, of course, material and human, but in addition science derives from society problems to be solved, ideas and metaphors to be used, techniques for inquiry, and, not least, standards of explanation (the notion of what will be accepted as an explanation in one time is so different from that which motivates thinking at another). Understanding the history of these external relations of science is essential to understanding our own science of the 1970's. And the interpretations of that history are not always favorable. They link science with society in such a way that science amplifies the social signals which stimulate it, and even exaggerates the worst of them.

One theorist, Husserl, understood modern science as a human 'project,' an attempt to treat Nature as an instrument, as an idealized extension of the utilization of prescientific craft tools. And this mathematical projection upon Nature, not only in Husserl's eyes, has evidently been successful. One might claim that the metaphor of Nature study as craft became literally true, for the human environment has been quite objectively changed. Nature can be, and has been, amenable to this projection. The mechanical may not be the only way to truth about Nature, but it has succeeded beyond expectation, beyond science in classic Greek society wherein there was curiosity and high intelligence but no drive for mastery over Nature and its processes.

To other thinkers, science shares the competitive and calculating spirit of modern society because it has been conceptualized in full parallel with the technological demands of industry. If so, science in its recent ways of thought and achievement will probably be judged less than universal in its scope and technique, less apt for those tasks of mankind which lie *outside* of, or which arise *after* full industrialization. But, how can one claim that science is restrictively characterized by the local epoch in which we now live? Briefly we can simply list the criteria for the empirical and theoretical success of science and see that they may be stated as criteria for a working, successful, well-engineered mass-production factory: *precision; simplicity; analysis of parts and components; impersonal, objective* and completely standardized workers and supplies; *economy* of thought, tools and materials; *efficiency* of administration and labor; *unified*, consistent, harmonious and *complete* development from raw materials to finished product; and finally *determinate* relations between input and output.

Any theory for man today, any philosophy of science, would reflect upon these emphasized criteria in terms which in turn reflect the industrial foundations of our times.

This relation of industrial society to its technological culture, and in particular to the scientific ways of thinking, may be seen by yet another comparison. Thus, Simmel and Schumpeter have investigated how the modern money economy, as it developed out of feudal society, *reflected* and, more importantly, *promoted* a scientific manner of thought as efficiently as did the developing technology; indeed I think the economic preceded the technological in this relationship by nearly three centuries. A money economy encourages such scientific thought by providing a social context wherein the standards are *abstract, impersonal, objective, quantitative, rational*, and where money itself is *conserved* and as empty of intrinsic perceived properties as physical mass in the particles and structures of Newtonian mechanics was sometimes thought to be.

The question of the social connections of thought is subtle, to be sure, and we cannot be certain about much in the history and sociology of ideas. If science amplified and was amplified by its society as I have just suggested, even receiving part of its rational methodology from that society, it must also be plausible that our science received whatever may be irrational in this society. However, what is irrational in modern technology and in a money economy has not often been agreed upon. The complex configuration of technology-money-science, taken as the material base and background for Western history, produced a technological civilization which has always been partly dehumanized. And within this partially dehumanized society, we see a situation where working and living are generally divorced, where the culture and enlightenment of individual persons is often and easily manipulated and, in the extreme, socially coordinated, where the concept and consciousness of happiness is itself regulated.

But I believe an optimistic note may be sounded. We have, as yet, a society of only *incomplete* technology. *Complete* technology, the total use of precise and automatic scientific achievements, would be a precondition not necessarily, as the gloomy humanists often fear, of an extrapolation of the present human situation, namely a totally dehumanized society, but rather of its opposite: thorough and complete utilization of science and technology would be the necessary *precondition* of a humanized society founded upon dehumanized production. Necessary, of course; surely not sufficient. But such completed science will compel man to make certain moral and political decisions, for it will be true that freedom from necessary and dehumanizing labor, if taken as the character of a civilization and not just of a leisure class, poses fundamental problems of knowledge and morality afresh. If our science cannot actually lead us to new ethical 'experiments in living' it will be the new tradition from which the possibility of such experiments will grow. We are not faced with the single road to an end to a period of history, a new decline and fall, but rather with another option, a transition which is fashioning its own way.

Unfortunately, it is plausible to expect that our science, here and now, will continue to function as an ideology as well as a servant of technology, aside from its success as conqueror of greater ranges of Nature. Long connected with elites, and now accustomed to and requiring support by political powers, science alone can scarcely be asked to infuse our socialist, capitalist, or mixed societies with a humanistic spirit. Weak as the model may be, we are fortunate enough that its internal ethic is a humane one. Only as the habits of thought which arose in the millenia of necessary work begin to be supplanted in social and personal life, might we also expect that new patterns of posing problems, new ways of thought, new ways of curiosity, new feelings for working and playing, might also arise within a new way of science which is open to all men. The leisure we cherish, and yet sometimes we fear, will bring, I suspect, this new way of knowing. For we will have to pass, in the next social transformation, from the essential prerequisite for all previous struggles for life, a *science of work*, to another prerequisite, a *science of pleasure*. And if this phrase is too puzzling, if it seems to demand that esthetics become science, and science learn much from art, at least it suggests the quality of the transition which a peaceful world would have ahead.

We do not know under what conditions the social impact of science would be wholly humane. But we can say that science provides one great quality of the enlightened and humane life: objectivity. And why objectivity? Surely we have met those who criticize science as coldly objective? But if we wish to have a society of citizens without illusion, self-governing, self-evaluating, judging for themselves what is possible, what is impossible, what is probable, and how to choose the most likely course in the light of inadequate evidence—and that life of probabilities and uncertainties is the scientific life—then we must see objectivity taught throughout our culture. Objectivity, after all, is the ability to see the facts; it never need entail an inability to assess their human significance.

And in the realm of ethics, science can contribute this utterly valuable core of objectivity in a number of ways:

1. logical necessity which should be inherent in the procedures of validating hypotheses.
2. logical consistency among the ethical norms of a proposed system.
3. factual objectivity of the characterization given of the empirical features of human attitudes and conduct, which are the subject-matter for moral appraisal.
4. factual objectivity of statements about the relations of means to ends, and of conditions to consequences.
5. factual objectivity of statements about human needs, interests, and ideals as they have arisen in social context.
6. conformity of the proposed norms of a system with the basic biological, social, and psychological nature of man, the preservation of his existence, the facts of growth, development, and evolution.
7. factual objectivity of the claims of universality which certain moral norms may embody, as shown in the conscience of men in certain groups, and perhaps invariantly across all groups.

In summary, developments in science relate to ethics as follows: first, scientific discoveries may force ethical decisions; second, scientific discoveries may make certain ethical decisions possible; third, scientific methods may help men to rational control and hence ethical planning, of their lives and societies; fourth, science may offer a model of democratic living for those who may be persuaded to see.

But, last, science should be modest. It is part of our own times and also part of our own weaknesses, not an escape from them. And it must not be a decision-making device for us. When choices are difficult, it is rational to choose in the light of probability estimates, rational to avoid the intellectual cowardice of Hamlet in his demand for factual and ethical certainty. As Reichenbach reminded us in his analysis of Hamlet's great soliloquy,[8] it takes more courage than *a priori* metaphysics knows in order to live by scientific criteria, that is, by empirical *and* rational standards. But we must never think that science gives us that moral courage. On the contrary, it is a moral choice to be courageous, an ethical act to be scientific.

Bibliography

1. See the pioneer study by Hans Kelsen, *Kausalität und Vergeltung* (translated into English as *Society and Nature*).
2. As recently as forty years ago, Reinhold Niebuhr entitled his somber and radical critique of modern social order *Moral Man and Immoral Society*.
3. See, Herbert Marcuse, *Reason and Revolution: Hegel and the Rise of Social Theory* (New York: Humanities Press, Inc.), especially pp. 16–28 and 253–257: also R.S. Cohen, "Dialectical Materialism and Carnap's Logical Empiricism," in *The Philosophy of Rudolph Carnap,* (La Salle, Ill.: Open Court Publishing Co.), ed. P.A. Schilpp, pp. 99–158.
4. John Wisdom, "Logical Constructions," (*Mind,* vol. 42, 1933) p. 195.
5. C.L. Stevenson, *Facts and Values* (New Haven: Yale Univ. Press, 1963) and earlier *Ethics and Language* (New Haven: Yale Univ. Press, 1944(.
6. Much of this and other paragraphs is derived from Abraham Edel's stimulating *Method in Ethical Theory* (Indianapolis: Bobbs-Merrill, 1963) and *Science and the Structure of Ethics* (Chicago: Univ. of Chicago Press, 1961).
7. Henri Chambre, *Le Marxism en Union Sovietique,* p. 334.
8. See Hans Reichenback *The Rise of Scientific Philosophy* (Berkeley: Univ. of California Press, 1951), Chapter 15.

THEODOSIUS DOBZHANSKY

Man and Natural Selection

Theodosius Dobzhansky is a zoologist and member and professor of the Rockefeller Institute. This article raises the important ethical questions involved in the "genetic management" of the human species. What, for example, will be the effect on the genetic pool of the future, of passing on "bad" genes, as medicine's ability to control gene-based diseases increases and as the bearers of these "bad" genes propagate? Should we have a program of "human eugenics,"

"Man and Natural Selection" by Theodosius Dobzhansky. In: *Science in Progress,* Thirteenth Series, edited by Wallace R. Brode. New Haven: Yale University Press, 1963, pp. 146–153. Copyright ©1963 by Yale University.

and if so, what methods should be applied, and again by whom and on the basis of which values?

Are Culture and Natural Selection Compatible?

. . . Several forms of natural selection operate in modern mankind. But they certainly do not operate as they did during the Stone Age or even as they did a century ago. Neither does natural selection operate always in the same way in wild and "natural" species, quite "unspoiled" by culture. This is inevitable. Natural selection depends on environments, and environments change. Human environments have changed a great deal in a century, not to speak of millennia.

The real problem is not whether natural selection in man is going on, but whether it is going on toward what we, humans, regard as betterment or deterioration. Natural selection tends to enhance the reproductive proficiency of the population in which it operates. Such proficiency is, however, not the only estimable quality with which we wish to see people endowed. And besides, a high reproductive fitness in one environment does not even insure the survival of the population or the species when the environment changes.

Normalizing selection is, as we have seen, not the only form of natural selection; the relaxation of some of its functions is, however, a cause for apprehension. Medicine, hygiene, civilized living save many lives which would otherwise be extinguished. This situation is here to stay; we would not want it to be otherwise, even if we could. Some of the lives thus saved will, however, engender lives that will stand in need of being saved in the generations to come. Can it be that we help the ailing, the lame, and the deformed only to make our descendants more ailing, more lame, and more deformed?

Suppose that we have learned how to save the lives of persons afflicted with a hereditary disease, such as retinoblastoma, which previously was incurably fatal. In genetic terms, this means that the Darwinian fitness of the victims of the disease has increased, and that the normalizing selection against this disease is relaxed. What will be the consequence? The incidence of the disease in the population will increase from generation to generation. The increase is likely to be slow, generally no more than by one mutation rate per generation. It may take centuries or millennia to notice the difference for any one disease or malformation, but the average health and welfare of the population are liable to show adverse effects of relaxed selection much sooner.

The process of mutation injects into every generation a certain number of harmful genes in the gene pool of the population; the process of normalizing selection eliminates a certain number of these genes. With environment reasonably stable, the situation tends to reach a state of equilibrium. At equilibrium, the mutation and the elimination are equal. If mutation becomes more frequent (as it does in man because of exposure to high-energy radiations and perhaps to some chemicals), or if the elimination is lagging because of relaxation of normalizing selection, the incidence of harmful mutant genes in the population is bound to increase. And take note of this: if the classical theory of population structure were correct, all harmful mutations would be in a sense equivalent.

For at equilibrium there is one elimination for every mutation, regardless of whether the mutation causes a lethal hereditary disease like retinoblastoma, or a malformation like achondroplasia, or a relatively mild defect such as myopia.

It would no doubt be desirable to eliminate from human populations all harmful mutant genes and to substitute for them favorable genes. But how is this end to be attained? A program of eugenics to achieve genetic health and eventual improvement of the human species has, in recent years, been urged with great eloquence, particularly by Muller, and many other authors: The fortunate few who happen to carry mostly "normal" or favorable genes should be better progenitors of the coming generations than are those who carry average, or heavier than average, genetic loads. Let us then take the semen of the superior males, and use it to produce numerous progeny by artificial insemination of women who will be happy to be mothers of children of the superior sires. Techniques will eventually be invented to obtain also the egg cells of superior females; indeed, the ovaries of human females are capable of producing numerous egg cells, most of which are at present wasted. It will then be possible to combine the finest egg cells with choicest sperms; the uteri of women who happen to be carriers of average or higher-than-average genetic loads will be good enough for the development of the genetically superior fetuses. But not even this would guarantee the best possible genetic endowments in the progeny. Very distinguished parents sometimes produce commonplace, and even inferior, children. The distant vista envisaged by Muller is tissue culture of (diploid) body cells of the very best donors, and a technique to stimulate these cells to develop without fertilization (parthenogenetically), thus giving rise to numerous individuals, all as similar to the donor and to each other as identical twins.

It may be doubted, however, whether modern genetics has progressed far enough to embark on a program as far-reaching as Muller suggests. Wright considers that the situation calls rather for research in what he describes neatly as "unfortunately the unpopular and scientifically somewhat unrewarding borderline fields of genetics and the social sciences." Although at equilibrium there may be one genetic elimination for every mutation, it is unrealistic to equate the human and social consequences of different mutations. The elimination of a lethal mutant which causes death of an embryo before implantation in the uterus is scarcely noticed by the mother or by anyone else. Suffering accompanies the elimination of a mutant, such as retinoblastoma, which kills an infant apparently normal at birth. Many mutants, such as hemophilia or Huntington's chorea, kill children, adolescents, or adults, cause misery to their victims, and disruption of the lives of their families. There is no way to measure precisely the amount of human anguish; yet one may surmise that the painful and slow death of the victims of so many hereditary diseases is torment greater than that involved in the elimination of a gene for achondroplasia owing to the failure of an achondroplastic dwarf to beget children.

Looked at from the angle of the costs to the society, the nonequivalence of different mutants is no less evident. Myopia may be inherited as a recessive trait. Increases of the frequency in populations of the gene for myopia are undesirable. Yet it may become more and more common in future generations. However, only a fanatic might advocate sterilization of the myopics or other radical measures to prevent the spread of this gene. One may hope that civilized societies can tolerate some more myopics; many of them are very useful citizens, and their defect can rather easily be corrected by a relatively inexpensive environmental change—wearing glasses. The effort needed to eradicate or to reduce the frequency of myopia genetically would exceed that requisite to rectify their defect environmentally, by manufacturing more pairs of glasses.

Diabetes mellitus is, given the present level of medicine, more difficult and expensive to correct than is myopia. Some diabetics may nevertheless be treated successfully by insulin therapy, helped to live to old age, and enabled to raise families as large as nondiabetics. The incidence of diabetes may therefore creep up slowly in the generations to come. Now, most people would probably agree that it is better to be free of diabetes than to have it under control, no matter how successfully, by insulin therapy or other means. The prospect is not a pleasant one to contemplate. Insulin injections may perhaps be almost as common in some remote future as taking aspirin tablets is at present.

Toward Guidance of Human Evolution

We are faced, then, with a dilemma—if we enable the weak and the deformed to live and to propagate their kind, we face the prospect of a genetic twilight; but if we let them die or suffer when we can save them we face the certainty of a moral twilight. How to escape this dilemma?

I can well understand the impatience which some of my readers may feel if I refuse to provide an unambiguous answer to so pressing a problem. Let me plead with you, however, that infatuation with oversimple answers to very complex and difficult problems is one of the earmarks of intellectual mediocrity. I am afraid that the problem of guidance of human evolution has no simple solution. At least I have not found one, nor has anybody else in my opinion. Each genetic condition will have to be considered on its own merits, and the solutions that may be adopted for different conditions will probably be different. Suppose that everybody agrees that the genes causing myopia, achondroplasia, diabetes, and retinoblastoma are undesirable. We shall nevertheless be forced to treat them differently. Some genetic defects will have to be put up with and managed environmentally; others will have to be treated genetically, by artificial selection, and the eugenic measures that may be needed can be effected without accepting any kind of biological Brave New World.

Let us face this fact: Our lives depend on civilization and technology, and the lives of our descendants will be even more dependent on civilized environments. I can imagine a wise old ape-man who deplored the softness of his contemporaries who used stone knives to carve their meat instead of doing this with their teeth; or a solid conservative Peking man viewing with alarm the newfangled habit of using fire to make oneself warm. I have yet to hear anyone seriously proposing that we give up the use of knives and fire now. Nor does anyone in his right mind urge that we let people die of smallpox or tuberculosis, in order that genetic resistance to these diseases be maintained. The remedy for our genetic dependence on technology and medicine is more, not less, technology and medicine. You may, if you wish, feel nostalgic for the good old days of our cave-dwelling ancestors; the point of no return was passed in the evolution of our species many millennia before anyone could know what was happening.

Of course, not all genetic defects can be corrected by tools or remedies or medicines. Even though new and better tools and medicines will, one may hope, be invented in the future, this will not make all genetic equipments equally desirable. It is a relatively simple matter to correct for lack of genetic resistance to smallpox by vaccination, or for myopia by suitable glasses. It is not so simple with many other genetic defects. Surgical removal of the eyes is called for in cases of retinoblastoma; this saves the lives of the victims, but leaves them blind. No remedies are known for countless other genetic defects. Human life is sacred; yet the social costs of some genetic variants are so great, and their social contributions are so small, that avoidance of their birth is ethically the most acceptable as well as the wisest solution. This does not necessarily call for enactment of Draconian eugenic laws; it is perhaps not overoptimistic to hope that spreading biological education and understanding may be a real help. Make persons whose progeny is likely to inherit a serious genetic defect aware of this fact; they may draw the conclusions themselves.

The strides accomplished by biochemical genetics in recent years have led some biologists to hope that methods will soon be discovered to induce specific changes in human genes of our choice. This would indeed be a radical solution of the problem of management of the evolution of our species and of other species as well. We would simply change the genes which we do not like, in ways conforming to our desires. Now, if the history of science has any lesson to teach us, it is the unwisdom of declaring certain goals to be unattainable. The cavalier way in which the progress of science often treats such predictions should instill due humility even in the most doctrinaire prophets. The best that can be said about the possibility of changing specific genes in man in accordance with our desires is that, although such an invention would be a great boon, it is not within reach yet. And it cannot be assumed to be achievable.

Let us also not exaggerate the urgency of the problem of the genetic management of the evolution of our species. Another problem, that of the runaway overpopulation of our planet, is far more immediate and critical. If mankind will prove unable to save itself from being choked by crowding, it hardly needs to worry about its genetic quality. Although the problems of numbers and of quality are not one and the same, they may yet be closely connected in practice. As steps toward regulation of the size of population begin to be taken, and this surely cannot be postponed much longer, the genetic problem will inexorably obtrude itself. The questions, "how many people" and "what kind of people" will be solved together, if they will be solved at all.

Some people believe that all would be well with mankind if only natural selection were permitted to operate without obstruction by medicine and technology. Let us not forget, however, that countless biological species of the past have become extinct, although their evolution was directed by natural selection unadulterated by culture. What we want is not simply natural selection, but selection, natural and artificial, directed toward humanly desirable goals. What are these goals? This is the central problem of human ethics and of human evolution. Darwinian fitness is no guide here. If, in some human society, genetically duller people produce more progeny than the brighter ones, this simply means that, in the environment of that particular society, being a bit thickheaded increases the Darwinian fitness, and being too intelligent decreases it. Natural selection will act accordingly, and will not be any less "natural" on that account.

Human cultural evolution has resulted in the formation of a system of values, of *human* values. These are the values to which we wish human evolution to conform. These values are products of cultural evolution, conditioned of course by the biological evolution, yet not deducible from the latter. Where do we find a criterion by which these values are to be judged? I know of no better one than that proposed by the ancient Chinese sage: "Every system of moral laws must be based upon man's own consciousness, verified by the common experience of mankind, tested by due sanction of historical experience and found without error, applied to the operations and processes of nature in the physical universe and found to be without contradiction, laid before the gods without question or fear, and able to wait a hundred generations and have it confirmed without a doubt by a Sage of posterity."

LEON R. KASS

The New Biology:
What Price Relieving Man's Estate?

Leon R. Kass was the executive secretary of the Committee on Life Sciences and Social Policy, National Research Council, National Academy of Sciences, Washington, D.C., and is presently a tutor at St. John's College, Annapolis, Maryland. This article by Kass logically follows the preceding one by Dobzhansky. Kass also addresses himself to the social and ethical problems raised by revolutionary advances in the biological sciences, in addition to the important but often neglected questions, "What is a good man, what is a good life for man, and what is a good community?"

This article is adapted from a working paper prepared by the author for the Committee on Life Sciences and Social Policy, as well as from lectures given at St. John's College in Annapolis, Maryland; at Oak Ridge National Laboratory, biology division, Oak Ridge, Tennessee; and at a meeting in Washington of the Council for the Advancement of Science Writing.

Leon R. Kass, "The New Biology: What Price Relieving Man's Estate?" *Science* 174:779–788, November 19, 1971. Copyright ©1971 by The American Association for the Advancement of Science. Reprinted with the permission of the author and publisher.

Recent advances in biology and medicine suggest that we may be rapidly acquiring the power to modify and control the capacities and activities of men by direct intervention and manipulation of their bodies and minds. Certain means are already in use or at hand, others await the solution of relatively minor technical problems, while yet others, those offering perhaps the most precise kind of control, depend upon further basic research. Biologists who have considered these matters disagree on the question of how much how soon, but all agree that the power for "human engineering," to borrow from the jargon, is coming and that it will probably have profound social consequences.

These developments have been viewed both with enthusiasm and with alarm; they are only just beginning to receive serious attention. Several biologists have undertaken to inform the public about the technical possibilities, present and future. Practitioners of social science "futurology" are attempting to predict and describe the likely social consequences of and public responses to the new technologies. Lawyers and legislators are exploring institutional innovations for assessing new technologies. All of these activities are based upon the hope that we can harness the new technology of man for the betterment of mankind.

Yet this commendable aspiration points to another set of questions, which are, in my view, sorely neglected—questions that inquire into the meaning of phrases such as the "betterment of mankind." A *full* understanding of the new technology of man requires an exploration of ends, values, standards. What ends will or should the new techniques serve? What values should guide society's adjustments? By what standards should the assessment agencies assess? Behind these questions lie others: what is a good man, what is a good life for man, what is a good community? This article is an attempt to provoke discussion of these neglected and important questions.

While these questions about ends and ultimate ends are never unimportant or irrelevant, they have rarely been more important or more relevant. That this is so can be seen once we recognize that we are dealing here with a group of technologies that are in a decisive respect unique: the object upon which they operate is man himself. The technologies of energy or food production, of communication, of manufacture, and of motion greatly alter the implements available to man and the conditions in which he uses them. In contrast, the biomedical technology works to change the user himself. To be sure, the printing press, the automobile, the television, and the jet airplane have greatly altered the conditions under which and the way in which men live; but men as biological beings have remained largely unchanged. They have been, and remain, able to accept or reject, to use and abuse these technologies; they choose, whether wisely or foolishly, the ends to which these technologies are means. Biomedical technology may make it possible to change the inherent capacity for choice itself. Indeed, both those who welcome and those who fear the advent of "human engineering" ground their hopes and fears in the same prospect: *that man can for the first time recreate himself.*

Engineering the engineer seems to differ in kind from engineering his engine. Some have argued, however, that biomedical engineering does not differ qualitatively from toilet training, education, and moral teachings—all of which are forms of so-called "social engineering," which has man as its object, and is used by one generation to mold the next. In reply, it must at least be said that the techniques which have hitherto been employed are feeble and inefficient when compared to those on the horizon. This quantitative difference rests in part on a qualitative difference in the means of intervention. The traditional influences operate by speech or by symbolic deeds. They pay tribute to man as the animal who lives by speech and who understands the meanings of actions. Also, their effects are, in general, reversible, or at least subject to attempts at reversal. Each person has greater or lesser power to accept or reject or abandon them. In contrast, biomedical engineering circumvents the human context of speech and meaning, bypasses choice, and goes directly to work to modify the human material itself. Moreover, the changes wrought may be irreversible.

In addition, there is an important practical reason for considering the biomedical technology apart from other technologies. The advances we shall examine are fruits of a large, humane project dedicated to the conquest of disease and the relief of human suffering. The biologist and physician, regardless of their private motives, are seen, with justification, to be the well-wishers and benefactors of mankind. Thus, in a time in which technological advance is more carefully scrutinized and increasingly criticized, biomedical developments are still viewed by most people as benefits largely without qualification. The price we pay for these developments is thus more likely to go unrecognized. For this reason, I shall consider only the dangers and costs of biomedical advance. As the benefits are well known, there is no need to dwell upon them here. My discussion is deliberately partial.

I begin with a survey of the pertinent technologies. Next, I will consider some of the basic ethical and social problems in the use of these technologies. Then, I will briefly raise some fundamental questions to which these problems point. Finally, I shall offer some very general reflections on what is to be done.

The Biomedical Technologies

The biomedical technologies can be usefully organized into three groups, according to their major purpose: (i) control of death and life, (ii) control of human potentialities, and (iii) control of human achievement. The corresponding technologies are (i) medicine, especially the arts of prolonging life and of controlling reproduction, (ii) genetic engineering, and (iii) neurological and psychological manipulation. I shall briefly summarize each group of techniques.

1. *Control of death and life.* Previous medical triumphs have greatly increased average life expectancy. Yet other developments, such as organ transplantation or replacement and research into aging, hold forth the promise of increasing not just the average, but also the maximum life expectancy. Indeed, medicine seems to be sharpening its tools to do battle with death itself, as if death were just one more disease.

More immediately and concretely, available techniques of prolonging life—respirators, cardiac pacemakers, artificial kidneys—are already in the lists against death. Ironically, the success of these devices in forestalling death has introduced confusion in determining that death has, in fact, occurred. The traditional signs of life—heartbeat and respiration—can now be maintained entirely by machines. Some physicians are now busily trying to devise so-called "new definitions of death," while others maintain that the technical advances show that death is not a concrete event at all, but rather a gradual process, like twilight, incapable of precise temporal localization.

The real challenge to death will come from research into aging and senescence, a field just entering puberty. Recent studies suggest that aging is a genetically controlled process, distinct from disease, but one that can be manipulated and altered by diet or drugs. Extrapolating from animal studies, some scientists have suggested that a decrease in the rate of aging might also be achieved simply by effecting a very small decrease in human body temperature. According to some estimates, by the year 2000 it may be technically possible to add from 20 to 40 useful years to the period of middle life.

Medicine's success in extending life is already a major cause of excessive population growth: death control points to birth control. Although we are already technically competent, new techniques for lowering fertility and chemical agents for inducing abortion will greatly enhance our powers over conception and gestation. Problems of definition have been raised here as well. The need to determine when individuals acquire enforceable legal rights gives society an interest in the definition of human life and of the time when it begins. These matters are too familiar to need elaboration.

Technologies to conquer infertility proceed alongside those to promote it. The first successful laboratory fertilization of human egg by human sperm was reported in 1969.[1] In 1970, British scientists learned how to grow human embryos in the laboratory up to at least the blastocyst stage [that is, to the age of 1 week].[2] We may soon hear about the next stage, the successful reimplantation of such an embryo into a woman previously infertile because of oviduct disease. The development of an artificial placenta, now under investigation, will make possible full laboratory control of fertilization and gestation. In addition, sophisticated biochemical and cytological techniques of monitoring the "quality" of the fetus have been and are being developed and used. These developments not only give us more power over the generation of human life, but make it possible to manipulate and to modify the quality of the human material.

2. *Control of human potentialities.* Genetic engineering, when fully developed, will wield two powers not shared by ordinary medical practice. Medicine treats existing individuals and seeks to correct deviations from a norm of health. Genetic engineering, in contrast, will be able to make changes that can be transmitted to succeeding generations and will be able to create new capacities, and hence to establish new norms of health and fitness.

Nevertheless, one of the major interests in genetic manipulation is strictly medical: to develop treatments for individuals with inherited diseases. Genetic disease is prevalent and increasing, thanks partly to medical advances that enable those affected to survive and perpetuate their mutant genes. The hope is that normal copies of the appropriate gene, obtained biologically or synthesized chemically, can be introduced into defective individuals to correct their deficiencies. This *therapeutic* use of genetic technology appears to be far in the future. Moreover, there is some doubt that it will ever be practical, since the same end could be more easily achieved by transplanting cells or organs that could compensate for the missing or defective gene product.

Far less remote are technologies that could serve *eugenic* ends. Their development has been endorsed by those concerned about a general deterioration of the human gene pool and by others who believe that even an undeteriorated human gene pool needs upgrading. Artificial insemination with selected donors, the eugenic proposal of Herman Muller,[3] has been possible for several years because of the perfection of methods for long-term storage of human spermatozoa. The successful maturation of human oocytes in the laboratory and their subsequent fertilization now make it possible to select donors of ova as well. But a far more suitable technique for eugenic purposes will soon be upon us— namely, nuclear transplantation, or cloning. Bypassing the lottery of sexual recombination, nuclear transplantation permits the asexual reproduction or copying of an already developed individual. The nucleus of a mature but unfertilized egg is replaced by a nucleus obtained from a specialized cell of an adult organism or embryo (for example, a cell from the intestines or the skin). The egg with its transplanted nucleus develops as if it had been fertilized and, barring complications, will give rise to a normal adult organism. Since almost all the hereditary material (DNA) of a cell is contained within its nucleus, the renucleated egg and the individual into which it develops are genetically identical to the adult organism that was the source of the donor nucleus. Cloning could be used to produce sets of unlimited numbers of genetically identical individuals, each set derived from a single parent. Cloning has been successful in amphibians and is now being tried in mice; its extension to man merely requires the solution of certain technical problems.

Production of man-animal chimeras by the introduction of selected nonhuman material into developing human embryos is also expected. Fusion of human and nonhuman cells in tissue culture has already been achieved.

Other, less direct means for influencing the gene pool are already available, thanks to our increasing ability to identify and diagnose genetic diseases. Genetic counselors can now detect biochemically and cytologically a variety of severe genetic defects (for example, Mongolism, Tay-Sachs disease) while the fetus is still in utero. Since treatments are at present largely unavailable, diagnosis is often followed by abortion of the affected fetus. In the future, more sensitive tests will also permit the detection of heterozygote carriers, the unaffected individuals who carry but a single dose of a given deleterious gene. The eradication of a given genetic disease might then be attempted by aborting all such carriers. In fact, it was recently suggested that the fairly common disease cystic fibrosis could be completely eliminated over the next 40 years by screening all pregnancies and aborting the 17,000,000 unaffected fetuses that will carry a single gene for this disease. Such zealots need to be reminded of the consequences should each geneticist be allowed an equal assault on his favorite genetic disorder, given that each human being is a carrier for some four to eight such recessive, lethal genetic diseases.

3. *Control of human achievement.* Although human achievement depends at least in part upon genetic endowment, heredity determines only the material upon which experience and education impose the form. The limits of many capacities and powers of an individual are indeed genetically determined, but the nurturing and perfection of these capacities depend upon other influences. Neurological and psychological manipulation hold forth the promise of controlling the development of human capacities, particularly those long considered most distinctively human: speech, thought, choice, emotion, memory, and imagination.

These techniques are now in a rather primitive state because we understand so little about the brain and mind. Nevertheless, we have already seen the use of electrical stimulation of the human brain to produce sensations of intense pleasure and to control rage, the use of brain surgery (for example, frontal lobotomy) for the relief of severe anxiety, and the use of aversive conditioning with electric shock to treat sexual perversion. Operant-conditioning techniques are widely used, apparently with success, in schools and mental hospitals. The use of so-called consciousness-expanding and hallucinogenic drugs is widespread, to say nothing of tranquilizers and stimulants. We are promised drugs to modify memory, intelligence, libido, and aggressiveness.

The following passages from a recent book by Yale neurophysiologist José Delgado—a book instructively entitled *Physical Control of the Mind: Toward a Psychocivilized Society*—should serve to make this discussion more concrete. In the early 1950's, it was discovered that, with electrodes placed in certain discrete regions of their brains, animals would repeatedly and indefatigably press levers to stimulate their own brains, with obvious resultant enjoyment. Even starving animals preferred stimulating these so-called pleasure centers to eating. Delgado comments on the electrical stimulation of a similar center in a human subject.[4]

The patient reported a pleasant tingling sensation in the left side of her body 'from my face down to the bottom of my legs.' She started giggling and making funny comments, stating that she enjoyed the sensation 'very much.' Repetition of these stimulations made the patient more communicative and flirtatious, and she ended by openly expressing her desire to marry the therapist.

And one further quotation from Delgado.[4]

Leaving wires inside of a thinking brain may appear unpleasant or dangerous, but actually the many patients who have undergone this experience have not been concerned about the fact of being wired, nor have they felt any discomfort due to the presence of conductors in their heads. Some women have shown their feminine adaptability to circumstances by wearing attractive hats or wigs to conceal their electrical headgear, and many people have been able to enjoy a normal life as outpatients, returning to the clinic periodically for examination and stimulation. In a few cases in which contacts were located in pleasurable areas, patients have had the opportunity to stimulate their own brains by pressing the button of a portable instrument, and this procedure is reported to have therapeutic benefits.

It bears repeating that the sciences of neurophysiology and psychopharma-
cology are in their infancy. The techniques that are now available are crude,
imprecise, weak, and unpredictable, compared to those that may flow from a
more mature neurobiology.

Basic Ethical and Social Problems in the Use of Biomedical Technology

After this cursory review of the powers now and soon to be at our disposal,
I turn to the questions concerning the use of these powers. First, we must
recognize that questions of use of science and technology are always moral and
political questions, never simply technical ones. All private or public decisions
to develop or to use biomedical technology—and decisions *not* to do so—
inevitably contain judgments about value. This is true even if the values guiding
those decisions are not articulated or made clear, as indeed they often are not.
Secondly, the value judgments cannot be derived from biomedical science. This
is true even if scientists themselves make the decisions.

These important points are often overlooked for at least three reasons.

1. They are obscured by those who like to speak of "the control of nature by
   science." It is men who control, not that abstraction "science." Science may
   provide the means, but men choose the ends; the choice of ends comes from
   beyond science.
2. Introduction of new technologies often appears to be the result of no decision
   whatsoever, or of the culmination of decisions too small or unconscious to be
   recognized as such. What can be done is done. However, someone is deciding
   on the basis of some notions of desirability, no matter how selfserving or
   altruistic.
3. Desires to gain or keep money and power no doubt influence much of what
   happens, but these desires can also be formulated as reasons and then discussed
   and debated.

Insofar as our society has tried to deliberate about questions of use, how has
it done so? Pragmatists that we are, we prefer a utilitarian calculus: we weigh
"benefits" against "risks," and we weigh them for both the individual and
"society." We often ignore the fact that the very definitions of "a benefit" and
"a risk" are themselves based upon judgments about value. In the biomedical
areas just reviewed, the benefits are considered to be self-evident: prolongation
of life, control of fertility and of population size, treatment and prevention of
genetic disease, the reduction of anxiety and aggressiveness, and the enhancement
of memory, intelligence, and pleasure. The assessment of risk is, in general,
simply pragmatic—will the technique work effectively and reliably, how much
will it cost, will it do detectable bodily harm, and who will complain if we
proceed with development? As these questions are familiar and congenial,
there is no need to belabor them.

The very pragmatism that makes us sensitive to considerations of economic
cost often blinds us to the larger social costs exacted by biomedical advances.
For one thing, we seem to be unaware that we may not be able to maximize

all the benefits, that several of the goals we are promoting conflict with each other. On the one hand, we seek to control population growth by lowering fertility; on the other hand, we develop techniques to enable every infertile woman to bear a child. On the one hand, we try to extend the lives of individuals with genetic disease; on the other, we wish to eliminate deleterious genes from the human population. I am not urging that we resolve these conflicts in favor of one side or the other, but simply that we recognize that such conflicts exist. Once we do, we are more likely to appreciate that most "progress" is heavily paid for in terms not generally included in the simple utilitarian calculus.

To become sensitive to the larger costs of biomedical progress, we must attend to several serious ethical and social questions. I will briefly discuss three of them: (1) questions of distributive justice, (2) questions of the use and abuse of power, and (3) questions of self-degradation and dehumanization.

Distributive Justice

The introduction of any biomedical technology presents a new instance of an old problem—how to distribute scarce resources justly. We should assume that demand will usually exceed supply. Which people should receive a kidney transplant or an artificial heart? Who should get the benefits of genetic therapy or of brain stimulation? Is "first-come, first-served" the fairest principle? Or are certain people "more worthy," and if so, on what grounds?

It is unlikely that we will arrive at answers to these questions in the form of deliberate decisions. More likely, the problem of distribution will continue to be decided ad hoc and locally. If so, the consequence will probably be a sharp increase in the already far too great inequality of medical care. The extreme case will be longevity, which will probably be, at first, obtainable only at great expense. Who is likely to be able to buy it? Do conscience and prudence permit us to enlarge the gap between rich and poor, especially with respect to something as fundamental as life itself?

Questions of distributive justice also arise in the earlier decisions to acquire new knowledge and to develop new techniques. Personnel and facilities for medical research and treatment are scarce resources. Is the development of a new technology the best use of the limited resources, given current circumstances? How should we balance efforts aimed at prevention against those aimed at cure, or either of these against efforts to redesign the species? How should we balance the delivery of available levels of care against further basic research? More fundamentally, how should we balance efforts in biology and medicine against efforts to eliminate poverty, pollution, urban decay, discrimination, and poor education? This last question about distribution is perhaps the most profound. We should reflect upon the social consequences of seducing many of our brightest young people to spend their lives locating the biochemical defects in rare genetic diseases, while our more serious problems go begging. The current squeeze on money for research provides us with an opportunity to rethink and reorder our priorities.

Problems of distributive justice are frequently mentioned and discussed, but they are hard to resolve in a rational manner. We find them especially difficult because of the enormous range of conflicting values and interests that characterizes our pluralistic society. We cannot agree—unfortunately, we often do not even try to agree—on standards for just distribution. Rather, decisions tend to be made largely out of a clash of competing interests. Thus, regrettably, the question of how to distribute justly often gets reduced to who shall decide how to distribute. The question about justice has led us to the question about power.

Use and Abuse of Power

We have difficulty recognizing the problems of the exercise of power in the biomedical enterprise because of our delight with the wondrous fruits it has yielded. This is ironic because the notion of power is absolutely central to the modern conception of science. The ancients conceived of science as the *understanding* of nature, pursued for its own sake. We moderns view science as power, as *control* over nature; the conquest of nature "for the relief of man's estate" was the charge issued by Francis Bacon, one of the leading architects of the modern scientific project.[5]

Another source of difficulty is our fondness for speaking of the abstraction "Man." I suspect that we prefer to speak figuratively about "Man's power over Nature" because it obscures an unpleasant reality about human affairs. It is in fact particular men who wield power, not Man. What we really mean by "Man's power over Nature" is a power exercised by some men over other men, with a knowledge of nature as their instrument.

While applicable to technology in general, these reflections are especially pertinent to the technologies of human engineering, with which men deliberately exercise power over future generations. An excellent discussion of this question is found in *The Abolition of Man,* by C.S. Lewis.[6]

It is, of course, a commonplace to complain that men have hitherto used badly, and against their fellows, the powers that science has given them. But that is not the point I am trying to make. I am not speaking of particular corruptions and abuses which an increase of moral virtue would cure: I am considering what the thing called "Man's power over Nature" must always and essentially be. . . .

In reality, of course, if any one age really attains, by eugenics and scientific education, the power to make its descendants what it pleases, all men who live after it are the patients of that power. They are weaker, not stronger: for though we may have put wonderful machines in their hands, we have preordained how they are to use them. . . . The real picture is that of one dominant age . . . which resists all previous ages most successfully and dominates all subsequent ages most irresistibly, and thus is the real master of the human species. But even within this master generation (itself an infinitesimal minority of the species) the power will be exercised by a minority smaller still. Man's conquest of Nature, if the dreams of some scientific planners are realized, means the rule of a few hundreds of men over billions upon billions of men. There neither is nor can be any simple increase of power on Man's side. Each new power won *by* man is a power *over* man as well. Each advance leaves him weaker as well as stronger. In every victory, besides being the general who triumphs, he is also the prisoner who follows the triumphal car.

Please note that I am not yet speaking about the problem of the misuse or abuse of power. The point is rather that the power which grows is unavoidably the power of only some men, and that the number of powerful men decreases as power increases.

Specific problems of abuse and misuse of specific powers must not, however, be overlooked. Some have voiced the fear that the technologies of genetic engineering and behavior control, though developed for good purposes, will be put to evil uses. These fears are perhaps somewhat exaggerated, if only because biomedical technologies would add very little to our highly developed arsenal for mischief, destruction, and stultification. Nevertheless, any proposal for large-scale human engineering should make us wary. Consider a program of positive eugenics based upon the widespread practice of asexual reproduction. Who shall decide what constitutes a superior individual worthy of replication? Who shall decide which individuals may or must reproduce, and by which method? These are questions easily answered only for a tyrannical regime.

Concern about the use of power is equally necessary in the selection of means for desirable or agreed-upon ends. Consider the desired end of limiting population growth. An effective program of fertility control is likely to be coercive. Who should decide the choice of means? Will the program penalize "conscientious objectors"?

Serious problems arise simply from obtaining and disseminating information, as in the mass screening programs now being proposed for detection of genetic disease. For what kinds of disorders is compulsory screening justified? Who shall have access to the data obtained, and for what purposes? To whom does information about a person's genotype belong? In ordinary medical practice, the patient's privacy is protected by the doctor's adherence to the principle of confidentiality. What will protect his privacy under conditions of mass screening?

More than privacy is at stake if screening is undertaken to detect psychological or behavioral abnormalities. A recent proposal, tendered and supported high in government, called for the psychological testing of all 6-year-olds to detect future criminals and misfits. The proposal was rejected; current tests lack the requisite predictive powers. But will such a proposal be rejected if reliable tests become available? What if certain genetic disorders, diagnosable in childhood, can be shown to correlate with subsequent antisocial behavior? For what degree of correlation and for what kinds of behavior can mandatory screening be justified? What use should be made of the data? Might not the dissemination of the information itself undermine the individual's chance for a worthy life and contribute to his so-called antisocial tendencies?

Consider the seemingly harmless effort to redefine clinical death. If the need for organs for transplantation is the stimulus for redefining death, might not this concern influence the definition at the expense of the dying? One physician, in fact, refers in writing to the revised criteria for declaring a patient dead as a "new definition of heart donor eligibility".[7]

Problems of abuse of power arise even in the acquisition of basic knowledge. The securing of a voluntary and informed consent is an abiding problem in the use of human subjects in experimentation. Gross coercion and deception are now rarely a problem; the pressures are generally subtle, often related to an intrinsic power imbalance in favor of the experimentalist.

A special problem arises in experiments on or manipulations of the unborn. Here it is impossible to obtain the consent of the human subject. If the purpose of the intervention is therapeutic—to correct a known genetic abnormality, for example—consent can reasonably be implied. But can anyone ethically consent to nontherapeutic interventions in which parents or scientists work their wills or their eugenic visions on the child-to-be? Would not such manipulation represent in itself an abuse of power, independent of consequences?

There are many clinical situations which already permit, if not invite, the manipulative or arbitrary use of powers provided by biomedical technology: obtaining organs for transplantation, refusing to let a person die with dignity, giving genetic counselling to a frightened couple, recommending eugenic sterilization for a mental retardate, ordering electric shock for a homosexual. In each situation, there is an opportunity to violate the will of the patient or subject. Such opportunities have generally existed in medical practice, but the dangers are becoming increasingly serious. With the growing complexity of the technologies, the technician gains in authority, since he alone can understand what he is doing. The patient's lack of knowledge makes him deferential and often inhibits him from speaking up when he feels threatened. Physicians *are* sometimes troubled by their increasing power, yet they feel they cannot avoid its exercise. "Reluctantly," one commented to me, "we shall have to play God." With what guidance and to what ends I shall consider later. For the moment, I merely ask: "By whose authority?"

While these questions about power are pertinent and important, they are in one sense misleading. They imply an inherent conflict of purpose between physician and patient, between scientist and citizen. The discussion conjures up images of master and slave, of oppressor and oppressed. Yet it must be remembered that conflict of purpose is largely absent, especially with regard to general goals. To be sure, the purposes of medical scientists are not always the same as those of the subjects experimented on. Nevertheless, basic sponsors and partisans of biomedical technology are precisely those upon whom the technology will operate. The will of the scientist and physician is happily married to (rather, is the offspring of) the desire of all of us for better health, longer life, and peace of mind.

Most future biomedical technologies will probably be welcomed, as have those of the past. Their use will require little or no coercion. Some developments, such as pills to improve memory, control mood, or induce pleasure, are likely to need no promotion. Thus, even if we should escape from the dangers of coercive manipulation, we shall still face large problems posed by the voluntary use of biomedical technology, problems to which I now turn.

Voluntary Self-Degradation and Dehumanization

Modern opinion is sensitive to problems of restriction of freedom and abuse of power. Indeed, many hold that a man can be injured only by violating his will. But this view is much too narrow. It fails to recognize the great dangers we shall face in the use of biomedical technology, dangers that stem from an excess of freedom, from the uninhibited exercises of will. In my view, our greatest problem will increasingly be one of voluntary self-degradation, or willing dehumanization.

Certain desired and perfected medical technologies have already had some dehumanizing consequences. Improved methods of resuscitation have made possible heroic efforts to "save" the severely ill and injured. Yet these efforts are sometimes only partly successful; they may succeed in salvaging individuals with severe brain damage, capable of only a less-than-human, vegetating existence. Such patients, increasingly found in the intensive care units of university hospitals, have been denied a death with dignity. Families are forced to suffer seeing their loved ones so reduced, and are made to bear the burdens of a protracted death watch.

Even the ordinary methods of treating disease and prolonging life have impoverished the context in which men die. Fewer and fewer people die in the familiar surroundings of home or in the company of family and friends. At that time of life when there is perhaps the greatest need for human warmth and comfort, the dying patient is kept company by cardiac pacemakers and defibrillators, respirators, aspirators, oxygenators, catheters, and his intravenous drip.

But the loneliness is not confined to the dying patient in the hospital bed. Consider the increasing number of old people who are still alive, thanks to medical progress. As a group, the elderly are the most alienated members of our society. Not yet ready for the world of the dead, not deemed fit for the world of the living, they are shunted aside. More and more of them spend the extra years medicine has given them in "homes for senior citizens," in chronic hospitals, in nursing homes—waiting for the end. We have learned how to increase their years, but we have not learned how to help them enjoy their days. And yet, we bravely and relentlessly push back the frontiers against death.

Paradoxically, even the young and vigorous may be suffering because of medicine's success in removing death from their personal experience. Those born since penicillin represent the first generation ever to grow up without the experience or fear of probable unexpected death at an early age. They look around and see that virtually all of their friends are alive. A thoughtful physician, Eric Cassell, has remarked on this in "Death and the Physician"[8].

While the gift of time must surely be marked as a great blessing, the *perception* of time, as stretching out endlessly before us, is somewhat threatening. Many of us function best under deadlines, and tend to procrastinate when time limits are not set. . . . Thus, this unquestioned boon, the extension of life, and the removal of the threat of premature death, carries with it an unexpected anxiety: the anxiety of an unlimited future.

In the young, the sense of limitless time has apparently imparted not a feeling of limitless opportunity, but increased stress and anxiety, in addition to the anxiety which results from other modern freedoms: personal mobility, a wide range of occupational choice, and independence from the limitations of class and familial patterns of work. . . . A certain aimlessness (often ringed around with great social consciousness) characterizes discussions about their own aspirations. The future is endless, and their inner demands seem minimal. Although it may appear uncharitable to say so, they seem to be acting in a way best described as "childish"—particularly in their lack of a time sense. They behave as though there were no tomorrow, or as though the time limits imposed by the biological facts of life had become so vague for them as to be nonexistent.

Consider next the coming power over reproduction and genotype. We endorse the project that will enable us to control numbers and to treat individuals with genetic desease. But our desires outrun these defensible goals. Many would welcome the chance to become parents without the inconvenience of pregnancy; others would wish to know in advance the characteristics of their offspring (sex, height, eye color, intelligence); still others would wish to design these characteristics to suit their tastes. Some scientists have called for the use of the new technologies to assure the "quality" of all new babies.[9] As one obstetrician put it: "The business of obstetrics is to produce *optimum* babies." But the price to be paid for the "optimum baby" is the transfer of procreation from the home to the laboratory and its coincident transformation into manufacture. Increasing control over the product is purchased by the increasing depersonalization of the process. The complete depersonalization of procreation (possible with the development of an artificial placenta) shall be, in itself, seriously dehumanizing, no matter how optimum the product. It should not be forgotten that human procreation not only issues new human beings, but is itself a human activity.

Procreation is not simply an activity of the rational will. It is a more complete human activity precisely because it engages us bodily and spiritually, as well as rationally. Is there perhaps some wisdom in that mystery of nature which joins the pleasure of sex, the communication of love, and the desire for children in the very activity by which we continue the chain of human existence? Is not biological parenthood a built-in "mechanism," selected because it fosters and supports in parents an adequate concern for and commitment to their children? Would not the laboratory production of human beings no longer be *human* procreation? Could it keep human parenthood human?

The dehumanizing consequences of programmed reproduction extend beyond the mere acts and processes of life-giving. Transfer of procreation to the laboratory will no doubt weaken what is presently for many people the best remaining justification and support for the existence of marriage and the family. Sex is now comfortably at home outside of marriage; child-rearing is progressively being given over to the state, the schools, the mass media, and the child-care centers. Some have argued that the family, long the nursery of humanity, has outlived its usefulness. To be sure, laboratory and governmental alternatives might be designed for procreation and child-rearing, but at what cost?

This is not the place to conduct a full evaluation of the biological family.

Nevertheless, some of its important virtues are, nowadays, too often overlooked. The family is rapidly becoming the only institution in an increasingly impersonal world where each person is loved not for what he does or makes, but simply because he is. The family is also the institution where most of us, both as children and as parents, acquire a sense of continuity with the past and a sense of commitment to the future. Without the family, we would have little incentive to take an interest in anything after our own deaths. These observations suggest that the elimination of the family would weaken ties to past and future, and would throw us, even more than we are now, to the mercy of an impersonal, lonely present.

Neurobiology and psychobiology probe most directly into the distinctively human. The technological fruit of these sciences is likely to be both more tempting than Eve's apple and more "catastrophic" in its result.[10] One need only consider contemporary drug use to see what people are willing to risk or sacrifice for novel experiences, heightened perceptions, or just "kicks." The possibility of drug-induced, instant, and effortless gratification will be welcomed. Recall the possibilities of voluntary self-stimulation of the brain to reduce anxiety, to heighten pleasure, or to create visual and auditory sensations unavailable through the peripheral sense organs. Once these techniques are perfected and safe, is there much doubt that they will be desired, demanded, and used?

What ends will these techniques serve? Most likely, only the most elemental, those most tied to the bodily pleasures. What will happen to thought, to love, to friendship, to art, to judgment, to public-spiritedness in a society with a perfected technology of pleasure? What kinds of creatures will we become if we obtain our pleasure by drug or electrical stimulation without the usual kind of human efforts and frustrations? What kind of society will we have?

We need only consult Aldous Huxley's prophetic novel *Brave New World* for a likely answer to these questions. There we encounter a society dedicated to homogeneity and stability, administered by means of instant gratifications and peopled by creatures of human shape but of stunted humanity. They consume, fornicate, take "soma," and operate the machinery that makes it all possible. They do not read, write, think, love, or govern themselves. Creativity and curiosity, reason and passion, exist only in a rudimentary and mutilated form. In short, they are not men at all.

True, our techniques, like theirs, may in fact enable us to treat schizophrenia, to alleviate anxiety, to curb aggressiveness. We, like they, may indeed be able to save mankind from itself, but probably only at the cost of its humanness. In the end, the price of relieving man's estate might well be the abolition of man.[11]

There are, of course, many other routes leading to the abolition of man. There are many other and better known causes of dehumanization. Disease, starvation, mental retardation, slavery, and brutality—to name just a few—have long prevented many, if not most, people from living a fully human life. We should work to reduce and eventually to eliminate these evils. But the existence of these evils should not prevent us from appreciating that the use of the technology of man, uninformed by wisdom concerning proper human ends, and untempered by an appropriate humility and awe, can unwittingly render us all irreversibly less

than human. For, unlike the man reduced by disease or slavery, the people
dehumanized à la *Brave New World* are not miserable, do not know that they are
dehumanized, and, what is worse, would not care if they knew. They are, indeed,
happy slaves, with a slavish happiness.

Some Fundamental Questions

The practical problems of distributing scarce resources, of curbing the abuses of
power, and of preventing voluntary dehumanization point beyond themselves to
some large, enduring, and most difficult questions: the nature of justice and the
good community, the nature of man and the good for man. My appreciation
of the profundity of these questions and my own ignorance before them makes
me hesitant to say any more about them. Nevertheless, previous failures to find
a shortcut around them have led me to believe that these questions must be
faced if we are to have any hope of understanding where biology is taking us.
Therefore, I shall try to show in outline how I think some of the larger questions
arise from my discussion of dehumanization and self-degradation.

My remarks on dehumanization can hardly fail to arouse argument. It might be
said, correctly, that to speak about dehumanization presupposes a concept of
"the distinctively human." It might also be said, correctly, that to speak about
wisdom concerning proper human ends presupposes that such ends do in fact
exist and that they may be more or less accessible to human understanding,
or at least to rational inquiry. It is true that neither presupposition is at home in
modern thought.

The notion of the "distinctively human" has been seriously challenged by
modern scientists. Darwinists hold that man is, at least in origin, tied to the subhuman; his seeming distinctiveness is an illusion or, at most, not very important.
Biochemists and molecular biologists extend the challenge by blurring the
distinction between the living and the nonliving. The laws of physics and chemistry are found to be valid and are held to be sufficient for explaining biological
systems. Man is a collection of molecules, an accident on the stage of evolution,
endowed by chance with the power to change himself, but only along determined
lines.

Psychoanalysts have also debunked the "distinctly human." The essence of man is
seen to be located in those drives he shares with other animals—pursuit of pleasure
and avoidance of pain. The so-called "higher functions" are understood to be
servants of the more elementary, the more base. Any distinctiveness or "dignity"
that man has consists of his superior capacity for gratifying his animal needs.

The idea of "human good" fares no better. In the social sciences, historicists
and existentialists have helped drive this question underground. The former hold
all notions of human good to be culturally and historically bound, and hence
mutable. The latter hold that values are subjective: each man makes his own, and
ethics becomes simply the cataloging of personal tastes.

Such appear to be the prevailing opinions. Yet there is nothing novel about
reductionism, hedonism, and relativism; these are doctrines with which Socrates
contended. What is new is that these doctrines seem to be vindicated by scientific

advance. Not only do the scientific notions of nature and of man flower into verifiable predictions, but they yield marvelous fruit. The technological triumphs are held to validate their scientific foundations. Here, perhaps, is the most pernicious result of technological progress—more dehumanizing than any actual manipulation or technique, present or future. We are witnessing the erosion, perhaps the final erosion, of the idea of man as something splendid or divine, and its replacement with a view that sees man, no less than nature, as simply more raw material for manipulation and homogenization. Hence, our peculiar moral crisis. We are in turbulent seas without a landmark precisely because we adhere more and more to a view of nature and of man which both gives us enormous power and, at the same time, denies all possibility of standards to guide its use. Though well-equipped, we know not who we are nor where we are going. We are left to the accidents of our hasty, biased, and ephemeral judgments.

Let us not fail to note a painful irony: our conquest of nature has made us the slaves of blind chance. We triumph over nature's unpredictabilities only to subject ourselves to the still greater unpredictability of our capricious wills and our fickle opinions. That we have a method is no proof against our madness. Thus, engineering the engineer as well as the engine, we race our train we know not where.[12]

While the disastrous consequences of ethical nihilism are insufficient to refute it, they invite and make urgent a reinvestigation of the ancient and enduring questions of what is a proper life for a human being, what is a good community, and how are they achieved.[13] We must not be deterred from these questions simply because the best minds in human history have failed to settle them. Should we not rather be encouraged by the fact that they considered them to be the most important questions?

As I have hinted before, our ethical dilemma is caused by the victory of modern natural science with its nonteleological view of man. We ought therefore to re-examine with great care the modern notions of nature and of man, which undermine those earlier notions that provide a basis for ethics. If we consult our common experience, we are likely to discover some grounds for believing that the questions about man and human good are far from closed. Our common experience suggests many difficulties for the modern "scientific view of man." For example, this view fails to account for the concern for justice and freedom that appears to be characteristic of all human societies.[14] It also fails to account for or to explain the fact that men have speech and not merely voice, that men can choose and act and not merely move or react. It fails to explain why men engage in moral discourse, or, for that matter, why they speak at all. Finally, the "scientific view of man" cannot account for scientific inquiry itself, for why men seek to know. Might there not be something the matter with a knowledge of man that does not explain or take account of his most distinctive activities, aspirations, and concerns.[15]

Having gone this far, let me offer one suggestion as to where the difficulty might lie: in the modern understanding of knowledge. Since Bacon, as I have mentioned earlier, technology has increasingly come to be the basic justification for scientific inquiry. The end is power, not knowledge for its own sake. But power is not only the end. It is also an important *validation* of knowledge. One definitely knows that one knows only if one can make. Synthesis is held to be the ultimate proof of understanding.[16] A more radical formulation holds that one knows only what one makes: knowing *equals* making.

Yet therein lies a difficulty. If truth be the power to change or to make the object studied, then of what do we have knowledge? If there are no fixed realities, but only material upon which we may work our wills, will not "science" be merely the "knowledge" of the transient and the manipulatable? We might indeed have knowledge of the laws by which things change and the rules for their manipulation, but no knowledge of the things themselves. Can such a view of "science" yield any knowledge about the nature of man, or indeed, about the nature of anything? Our questions appear to lead back to the most basic of questions: What does it mean to know? What is it that is knowable?[17]

We have seen that the practical problems point toward and make urgent certain enduring, fundamental questions. Yet while pursuing these questions, we cannot afford to neglect the practical problems as such. Let us not forget Delgado and the "psychocivilized society." The philosophical inquiry could be rendered moot by our blind, confident efforts to dissect and redesign ourselves. While awaiting a reconstruction of theory, we must act as best we can.

What Is To Be Done?

First, we sorely need to recover some humility in the face of our awesome powers. The arguments I have presented should make apparent the folly of arrogance, of the presumption that we are wise enough to remake ourselves. Because we lack wisdom, caution is our urgent need. Or to put it another way, in the absence of that "ultimate wisdom," we can be wise enough to know that we are not wise enough. When we lack sufficient wisdom to do, wisdom consists in not doing. Caution, restraint, delay, abstention are what this second-best (and, perhaps, only) wisdom dictates with respect to the technology for human engineering.

If we can recognize that biomedical advances carry significant social costs, we may be willing to adopt a less permissive, more critical stance toward new developments. We need to reexamine our prejudice not only that all biomedical innovation is progress, but also that it is inevitable. Precedent certainly favors the view that what can be done will be done, but is this necessarily so? Ought we not to be suspicious when technologists speak of coming developments as automatic, not subject to human control? Is there not something contradictory in the notion that we have the power to control all the untoward consequences of a technology, but lack the power to determine whether it should be developed in the first place?

What will be the likely consequences of the perpetuation of our permissive and fatalistic attitude toward human engineering? How will the large decisions be made? Technocratically and self-servingly, if our experience with previous technologies is any guide. Under conditions of laissez-faire, most technologists will pursue techniques, and most private industries will pursue profits. We are fortunate that, apart from the drug manufacturers, there are at present in the biomedical area few large industries that influence public policy. Once these appear, the voice of "the public interest" will have to shout very loudly to be heard above their whisperings in the halls of Congress. These reflections point to the need for institutional controls.

Scientists understandably balk at the notion of the regulation of science and technology. Censorship is ugly and often based upon ignorant fear; bureaucratic regulation is often stupid and inefficient. Yet there is something disingenuous about a scientist who professes concern about the social consequences of science, but who responds to every suggestion of regulation with one or both of the following: "No restrictions on scientific research," and "Technological progress should not be curtailed." Surely, to suggest that *certain* technologies ought to be regulated or forestalled is not to call for the halt of *all* technological progress (and says nothing at all about basic research). Each development should be considered on its own merits. Although the dangers of regulation cannot be dismissed, who, for example, would still object to efforts to obtain an effective, complete, global prohibition on the development, testing, and use of biological and nuclear weapons?

The proponents of laissez-faire ignore two fundamental points. They ignore the fact that not to regulate is as much a policy decision as the opposite, and that it merely postpones the time of regulation. Controls will eventually be called for— as they are now being demanded to end environmental pollution. If attempts are not made early to detect and diminish the social costs of biomedical advances by intelligent institutional regulation, the society is likely to react later with more sweeping, immoderate, and throttling controls.

The proponents of laissez-faire also ignore the fact that much of technology is already regulated. The federal government is already deep in research and development (for example, space, electronics, and weapons) and is the principal sponsor of biomedical research. One may well question the wisdom of the direction given, but one would be wrong in arguing that technology cannot survive social control. Clearly, the question is not control versus no control, but rather what kind of control, when, by whom, and for what purpose.

Means for achieving international regulation and control need to be devised. Biomedical technology can be no nation's monopoly. The need for international agreements and supervision can readily be understood if we consider the likely American response to the successful asexual reproduction of 10,000 Mao Tse-tungs.

To repeat, the basic short-term need is caution. Practically, this means that we should shift the burden of proof to the *proponents* of a new biomedical technology. Concepts of "risk" and "cost" need to be broadened to include some of the social and ethical consequences discussed earlier. The probable or possible harmful effects of the widespread use of a new technique should be anticipated and introduced as "costs" to be weighed in deciding about the *first* use. The regulatory institutions should be encouraged to exercise restraint and to formulate the grounds for saying "no." We must all get used to the idea that biomedical technology makes possible many things we should never do.

But caution is not enough. Nor are clever institutional arrangements. Institutions can be little better than the people who make them work. However worthy our intentions, we are deficient in understanding. In the *long* run, our hope can only lie in education: in a public educated about the meanings and limits of science and enlightened in its use of technology; in scientists better educated to understand the relationships between science and technology on the one hand, and ethics and politics on the other; in human beings who are as wise in the latter as they are clever in the former.

Bibliography

1. R.G. Edwards, B.D. Bavister, P.C. Steptoe, *Nature,* vol. 221, 632 (1969).
2. R.G. Edwards, P.C. Steptoe, J.M. Purdy, *Nature,* vol. 227, 1307 (1970).
3. H.J. Muller, *Science,* vol. 134, 643 (1961).
4. J.M.R. Delgado, *Physical Control of the Mind: Toward a Psychocivilized Society* (New York: Harper & Row, 1969). p. 185–88.
5. F. Bacon, *The Advancement of Learning, Book I,* H.G. Dick, ed. (New York: Random House, 1955), p. 193.
6. C.S. Lewis, *The Abolition of Man* (New York: Macmillan, 1965), pp. 69–71.
7. D.D. Rutstein, *Daedalus* (Spring 1969), p. 526.
8. E.J. Cassell, *Commentary* (June 1969), p. 76.
9. B. Glass, *Science,* vol. 171, 23 (1971).
10. It is, of course, a long-debated question as to whether the fall of Adam and Eve ought to be considered "catastrophic," or more precisely, whether the Hebrew tradition considered it so. I do not mean here to be taking sides in this quarrel by my use of the term "catastrophic," and, in fact, tend to line up on the negative side of the questions, as put above. Curiously, as Aldous Huxley's *Brave New World* [(New York: Harper & Row, 1969)] suggests, the implicit goal of the biomedical technology could well be said to be the reversal of the Fall and a return of man to the hedonic and immortal existence of the Garden of Eden. Yet I can point to at least two problems. First, the new Garden of Eden will probably have no gardens; the received, splendid world of nature will be buried beneath asphalt, concrete, and other human fabrications, a transformation that is already far along. (Recall that in *Brave New World* elaborate consumption-oriented, mechanical amusement parks—featuring, for example, centrifugal bumble-puppy- had supplanted wilderness and even ordinary gardens.) Second, the new inhabitant of the new "Garden" will have to be a creature for whom we have no precedent, a creature as difficult to imagine as to bring into existence. He will have to be simultaneously an innocent like Adam and a technological wizard who keeps the "Garden" running. (I am indebted to Dean Robert Goldwin, St. John's College, for this last insight.)
11. Some scientists naively believe that an engineered increase in human intelligence will steer us in the right direction. Surely we have learned by now that intelligence, whatever it is and however measured, is not synonymous with wisdom and that, if harnessed to the wrong ends, it can cleverly perpetrate great folly and evil. Given the activities in which many, if not most, of our best minds are now engaged, we should not simply rejoice in the prospect of enhancing IQ. On what would this increased intelligence operate? At best, the programming of further increases in IQ. It would design and operate techniques for prolonging life, for engineering reproduction, for delivering gratifications. With no gain in wisdom, our gain in intelligence can only enhance the rate of our dehumanization.
12. The philosopher Hans Jonas has made the identical point: "Thus the slow-working accidents of nature, which by the very patience of their small increments, large numbers, and gradual decisions, may well cease to be 'accident' in outcome, are to be replaced by the fast-working accidents of man's hasty and biased decisions, not exposed to the long test of the ages. His uncertain ideas are to set the goals of generations, with a certainty

borrowed from the presumptive certainty of the means. The latter presumption is doubtful enough, but this doubtfulness becomes secondary to the prime question that arises when man indeed undertakes to 'make himself': in what image of his own devising shall he do so, even granted that he can be sure of the means? In fact, of course, he can be sure of neither, not of the end, nor of the means, once he enters the realm where he plays with the roots of life. Of one thing only can he be sure: of his power to move the foundations and to cause incalculable and irreversible consequences. Never was so much power coupled with so little guidance for its use." [*J. Cent. Cong. Amer. Rabbis* (January 1968), p. 27.] These remarks demonstrate that, contrary to popular belief, we are not even on the right road toward a rational understanding of and rational control over human nature and human life. It is indeed the height of irrationality triumphantly to pursue rationalized technique, while at the same time insisting that questions of ends, values, and purposes lie beyond rational discourse.

13. It is encouraging to note that these questions are seriously being raised in other quarters— for example, by persons concerned with the decay of cities or the pollution of nature. There is a growing dissatisfaction with ethical nihilism. In fact, its tenets are unwittingly abandoned, by even its staunchest adherents, in any discussion of "what to do." For example, in the biomedical area, everyone, including the most unreconstructed and technocratic reductionist, finds himself speaking about the use of powers for "human betterment." He has wandered unawares onto ethical ground. One cannot speak of "human betterment" without considering what is meant by *the human* and by the related notion of *the good for man.* These questions can be avoided only by asserting that practical matters reduce to tastes and power, and by confessing that the use of the phrase "human betterment" is a deception to cloak one's own will to power. In other words, these questions can be avoided only by ceasing to discuss.

14. Consider, for example, the widespread acceptance, in the legal systems of very different societies and cultures, of the principle and the practice of third-party adjudication of disputes. And consider why, although many societies have practiced slavery, no slave-holder has preferred his own enslavement to his own freedom. It would seem that some notions of justice and freedom, as well as right and truthfulness, are constitutive for any society, and that a concern for these values may be a fundamental characteristic of "human nature."

15. Scientists may, of course, continue to believe in righteousness or justice or truth, but these beliefs are not grounded in their "scientific knowledge" of man. They rest instead upon the receding wisdom of an earlier age.

16. This belief, silently shared by many contemporary biologists, has recently been given the following clear expression: "One of the acid tests of understanding an object is the ability to put it together from its component parts. Ultimately, molecular biologists will attempt to subject their understanding of all structure and function to this sort of test by trying to synthesize a cell. It is of some interest to see how close we are to this goal." [P. Handler, ed. *Biology and the Future of Man* (New York: Oxford Univ. Press, 1970), p. 55.]

17. When an earlier version of this article was presented publicly, it was criticized by one questioner as being "antiscientific." He suggested that my remarks "were the kind that gave science a bad name." He went on to argue that, far from being the enemy of morality, the pursuit of truth was itself a highly moral activity, perhaps the highest. The relation of science and morals is a long and difficult question with an illustrious history, and it deserves a more extensive discussion than space permits. However, because some readers may share the questioner's response, I offer a brief reply. First, on the matter of reputation, we should recall that the pursuit of truth may be in tension with keeping a good name (witness Oedipus, Socrates, Galileo, Spinoza, Solzhenitsyn). For most of human history, the pursuit of truth (including "science") was not a reputable activity among the many, and was, in fact, highly suspect. Even today, it is doubtful whether more than a few appreciate knowledge as an end in itself. Science has acquired a "good

name" in recent times largely because of its technological fruit; it is therefore to be expected that a disenchantment with technology will reflect badly upon science. Second, my own attack has not been directed against science, but against the use of *some* technologies and, even more, against the unexamined belief—indeed, I would say, superstition—that all biomedical technology is an unmixed blessing. I share the questioner's belief that the pursuit of truth is a highly moral activity. In fact, I am inviting him and others to join in a pursuit of the truth about whether all these new technologies are really good for us. This is a question that merits and is susceptible of serious intellectual inquiry. Finally, we must ask whether what we call "science" has a monopoly on the pursuit of truth. What is "truth"? What is knowable, and what does it mean to know? Surely, these are also questions that can be examined. Unless we do so, we shall remain ignorant about what "science" is and about what it discovers. Yet "science"—that is, modern natural science—cannot begin to answer them; they are philosophical questions, the very ones I am trying to raise at this point in the text.

SHELDON KRIMSKY

The Scientist As Alienated Man

Sheldon Krimsky is an Assistant Professor of Philosophy at the University of South Florida. His principal field of publication and research is the philosophy of physics. In this essay Krimsky brings the Marxian concept of alienation to bear on the plight of the contemporary scientist. He sees the scientist as being alienated because of forces operating in modern science in several different, but related, aspects including the need for increasing specialization, the lack of historical roots in science curricula, and the separation of the scientist from control over the products of his labor.

The term alienation has been used so loosely in recent years that it has been practically stripped of its historical significance. That young people are alienated from their parents suggests a kind of estrangement we have termed "generation gap." Students who are alienated from traditional academic goals are proclaiming that education is irrelevant to their real needs. Blacks who are alienated from white society are telling us that white society and, in particular, the white power structure is racist. Homosexuals are alienated from themselves when they attempt to conceal and struggle with honest feelings of self-identity. In some sense the word alienation means "alien to" or "apart from." In Marxian terms, however, alienation describes a more complex relationship than mental detachment, physical hardship, or psychological anguish.

When Karl Marx wrote about the alienation of man in his early manuscripts he focused his analysis on the industrial working class. It is, however, a tribute to the depth of his inquiry and the richness of his categories that diverse forms of

Special contribution to *Science, Technology, and Freedom*; previously unpublished.

human productive activity extending beyond the industrial proletariat may be understood more clearly. In this essay we discuss the roots and forms of alienation in science, that is the alienation between the contemporary scientist and his work, by appropriating some concepts from the classical Marxist analysis. Five areas are discussed: (1) Science as Method: Specialization and Estrangement, (2) Science Curriculum: Neglecting the Genesis of Its Content, (3) Appropriation of the Products of Science, (4) Devaluation of Science, and (5) The Overglut and Dehumanization.

Science as Method: Specialization and Estrangement

Within the esoteric passages of Hegel's *Phenomenology of Mind* we find the seeds of the Marxian concept of alienation. In Hegel's odyssey the development of human consciousness is dialectical—it follows a pattern of conflict and reconciliation. His model is primarily epistemological. The knowing subject and the object of knowledge are alien to one another. A resolution of each stage of the divided consciousness brings the mind to a higher stage. The process is then repeated.

Marx attacked Hegel's metaphysical structure and provided a materialistic interpretation of many of the Hegelian categories. The unhappy consciousness of Hegel, which saw its object as alien, became the sum total of the relationships of a worker to the object of his labor. For Marx the industrial laborer is alienated from his work because (1) he is separated from the object of his labor, and (2) the object of labor rather than serving as a reward contributes to his oppression.[1] The worker labors on and understands a very small part of the product to which he contributes. He doesn't see a plan from start to finish but merely sees a blueprint for a highly particularized task. The worker is also separated from the product of his labor because he has no control over it. The assembly line worker in an automobile plant has as little control over the design and materials that go into making a new car as he does safety standards. The days of the craftsman have all but disappeared. Even the local shoemaker is slowly vanishing while synthetic shoes and rubber-soled shoes, more profitable to discard than repair, are displacing the longer wear models. Industrial workers, who labor on products, relate to them in parts. And scientists are no different. The proliferation of scientific specializations in the twentieth century has kept its pace with the specialized tasks of industrial assembly line work.

Scientists are trained to relate to their discipline in segments. When I attended graduate school it was rare to find students in solid state physics attending colloquia in particle physics. The experimentalists kept company with experimentalists and likewise theorists spent time with their academic brethren. The productivity of research in this century, a consequence of an exponential increase in advanced technology and an expanded university system enabling many more people the opportunity to acquire a science education, has created disciplines out of a discipline. Every major area of science is divided into mini-disciplines each with its own society and specialized journal.[2] One is no longer simply a physicist but a molecular spectroscopist, a low temperature man or a plasma physicist, and so forth.

There is a growing feeling among many young scientists that specialization is not to their advantage. What happens if their field of specialization is no longer in vogue?[3] Suppose the government loses interest in certain areas of basic research. Highly specialized work can also be an obstacle for changing jobs.[4] With only several research centers involved in their specialty they are restricted to certain geographical centers. There is of course always college teaching. But many colleges shy away from narrowly trained Ph.D.'s who have little or no teaching experience.

At an orientation for newly admitted graduate students I was once told that by the time I received my doctorate I would know more about a minute area of science than anyone else in the world. That sounded exciting at first. What a marvelous thought to be that unique. But suppose no one else wishes to know about your area? Regardless of how important one's work might seem, of how educated one might be, without social recognition there is only despair.

It has been argued that science itself has not contributed to the alienation of the scientist. Science is, after all, only a method, "a way of looking at the world; a procedure of asking and answering questions, a body of data collected by orderly, controlled procedures."[5]

But the scientific method is analytical and that implies specialization. Consequently, the scientific estate has established a separation of duties. Specialization has been one of the contributing factors to alienation. Man feels alone and minuscule. His contribution to society seems inconsequential.[6] He is more easily manipulated when his specific job is not needed. Once someone has committed himself to a specialty, after having spent more than a decade in the university, he is generally stuck with it. His success is contingent upon a dedication to, and prolific achievement in, a narrow area of research.

Another effect of specialization is that it permits scientists to rationalize their immunity from moral culpability. The fact that a man's work is highly specialized more easily permits him to abdicate any responsibility for the misuses of the products of his craft. He argues that the evil lies in the total product while his contribution is a minor and insignificant part of that totality.[7]

Science Curriculum: Neglecting the Genesis of Its Content

A fundamental insight of black revolutionaries has been epistemological. Understanding of the present is blinded and incomplete without knowledge of the past. For this reason Black identity has meant rediscovering Black history. Alienation has meant being severed from one's roots. The analogy we wish to make concerns the education of the scientist. In most cases the scientist has spent many arduous years studying his field. But what exactly does he study? He studies science by poring through contemporary textbooks in specialized areas and then, of course, he advances to journals.

But the study of science through technical texts leaves the scientist ignorant of the history of his field. What we find in most textbooks are achievements. Any history, that is covered, is treated summarily in an introductory section. The historical exigesis is usually presented as a smooth transition period. One gets the impression that the author feels some obligation to provide an introductory synopsis of the history of the subject matter. Needless to say, it is a very select history of achievements, or shall we say advancements, since failure can be an achievement. The scientist receives a distorted picture of the historical development of his field. Thomas Kuhn sums it up this way:

Textbooks thus begin by truncating the scientists sense of his discipline's history and then proceed to supply a substitute for what they have eliminated. Characteristically, textbooks of science contain just a bit of history, either in an introductory chapter or, more often, in scattered references to great heroes of an earlier age. From such references both students and professionals come to feel like participants in a long standing historical tradition. Yet the textbook derived tradition in which scientists come to sense their participation is one that, in fact, never existed. . . . [8]

Since most contemporary science curricula do not include studies in the history of science the scientist becomes severed from the historical roots of his knowledge.[9]

Another growing concern of the alienating aspects of science education has been brought to our attention by C.P. Snow's "two culture controversy." Science programs are so specialized that they make it all but impossible for the student to pursue a classical humanities education. The result is a growing rift between the scientific estate and the humanistically educated literati. The wider the gap in educational curriculum—the analytic, vertically progressive, highly specialized and value deprived education for the scientist versus the synthetic, historical and value infused education of the humanist—the greater the difficulty in communication between the disparate classes. Without such communication each is deprived of a realistic idea of the totality which includes natural laws and human ideals.

Appropriation of the Products of Science

". . . The external character of labor for the worker appears in the fact that it is not his own, but someone else's, that it does not belong to him, that in it he belongs not to himself, but to another."[10]

One of the major forms of industrial alienation as outlined by Marx is that the product of the laborer does not belong to him—it belongs to the capitalist. The worker has no control over the quantity, quality or use to which the product is put. A worker for Dow Chemical Corporation may be in complete disagreement with a war in which his country is engaged but he is excluded from any decisions on whether the company should contribute to the war effort.

For the scientist, the product of his labor is his research and, concretely, the papers he produces. Having control over the product of his labor means, for the scientist, that he can send his manuscript to the journal of his choice for publication or decide not to publish it at all.

But the more government extends its tentacles into the scientific community, the less will scientists have control over the products of their labor. The government will decide if something is classified, if it is publishable, and when and where it is to be published. In other words, within the seemingly humanistic and pro-scientific vehicle called "government sponsored research" lie the roots of an alienation analogous to what one finds with the industrial proletariat, namely the appropriation of the products of labor.[11]

Scientists working in industry must also follow rigid regulations concerning what can and cannot be published. The reasons may be different from those for scientists under government contract, but the effect is the same.

The product of the scientist, once out of his hands, should be part of the public domain. If circumstances arise whereby some scientific result may, if publicized, be detrimental to the general welfare of a nation-state or to humanity as a whole, then those consequences can be discussed by a community of scientists and lay people representing the public interest. Such debates can be carried on without divulging the specific recipe for the research.

The Devaluation of Science

Two major factors have contributed to the devaluation of science in this century: (1) scientific theories are no longer considered to be our microscopes into ultimate reality but are more generally seen as conventions for ordering phenomena, and (2) there is a growing feeling that the theoretical results of science are closely tied to destructive technology. Consequently, both the epistemological and the instrumental values of science have suffered.

The revolutions in physics, at the turn of the century, produced a complete reconstruction of such basic classical concepts as space, time, and matter. If Aristotle's physics survived for 1800 years, only to be replaced by Newton's physics with a lifetime of over 200 years, how many years will it be before Einstein's and Bohr's views are surpassed? The point is that we can no longer accept any theory as being the final word—each theory is tentative and therefore epistemologically suspect. The conceptual revolutions in physics have also raised the question of whether science can provide an unambiguous picture of reality. According to its orthodox interpretation, quantum mechanics is dualistic; both wave and particle models complement one another to form the complete picture.[12] This has made it awkward to ask questions about the nature of reality. In some experimental arrangements the microscopic entity acts as if it were undulatory while in other experimental arrangements it behaves as if it were corpuscular. But what is it? Certainly something cannot be both A and not-A simultaneously. It can, however, appear to be A in circumstance C_1 and appear to be not-A in circumstance C_2. What is actually behind the appearances is beyond our grasp.[13]

All this has produced a generation of scientists who, when pressed, take on the cloak of conventionalism. The search for the true nature of reality becomes an extraneous and unattainable part of scientific research—to be identified with the pejorative term "metaphysics." What is important, is that the theories work.

Since theories are the products of the scientist's labors, those products have lost their intrinsic epistemological value.

It is not uncommon to hear, among scientists, the view that the social or instrumental value of pure science is neutral; science does, however, have an intrinsic value by virtue of the fact that it contributes to our knowledge of nature—an Aristotelian legacy that knowledge is a good in and of itself. Political institutions, it is argued, by appropriating the pure products of science for their technology, make it appear that theoretical science is value-linked to the technology. But the argument, that scientists are immune from moral culpability since it is their *neutral* research which can be used either for good or for evil, is losing its persuasive force. Along with the obvious implications of one's theoretical research and the overt connections between it and the motives of political institutions, there is frequently little time lag between the theoretical research and the inimical use to which it is put. There is, as well, the true moral dilemma experienced by Einstein when he was asked to support the atomic bomb program in the United States.[14] If we don't produce it first the enemy will destroy multitudes. When the choice is between two evils, at best, we can only do less wrong. But the anguish that many scientists feel toward their colleagues is not based upon moral deliberations but rather upon economic ones. Today one does not hear: "If I don't produce a defoliant for use in Vietnam, then the Vietnamese will defoliate our green forests," but rather: "If I don't produce a defoliant some other scientist will, so why should I miss out on the lucrative contract?"

Wherein lies the alienation? It lies in the value of one's product. First, conventionalistic attitudes toward science have stripped scientific theories of their truth value. Second, the great myth of value-free science has been destroyed. Many scientists no longer see the pure theory and the applied theory as value-independent. The technology which gives content to a theory confronts us as a destructive force, e.g., as in the possibility for political and social control through behavior modification, the effects of carrying on chemical and biological warfare, or the spectre of a nuclear holocaust.

The alienation of the worker in his product means not only that the labor becomes an object, an external existence, but that is exists outside him, independently, as something alien to him, and that it becomes a power on its own confronting him. It means that the life which he has conferred on the object confronts him as something hostile and alien.[15]

The Overglut and Dehumanization

In the 1970's, perhaps more than ever in America's history, scientists are experiencing what industrial laborers have endured for centuries—the indignity of not having a secure prospect for employment. A graduate could no longer be guaranteed of a job commensurate with his education. In the 50's and 60's there was a great demand for scientists of all types. Great expectations were felt for the decades to come. But things didn't work out so well. Even people who attended the country's most prestigious institutions of science and technology and dedicated themselves to highly specialized research found that jobs were unavailable when they graduated.[16] It is not uncommon for small nonresearch oriented colleges to be receiving unsolicited applications from many hundreds of young Ph.D.'s desperately searching for a place to teach and practice their craft. After spending between five and seven years at a prestigious university and discovering that there is no demand for your skills it is quite normal to become alienated from your work, but worse than that, from yourself. A breakdown in self-esteem and with it the desire to be some other is undoubtedly the most heightened form of alienation.

The problem of job security felt by industrial workers in the late 1800's, a consequence of mechanization and large-scale immigration, was one of the motivating factors for unionization. But scientists are feeling the effect of the glut of academics and the unionization of professional scientists is being widely discussed.[17] Unlike past decades when recruiters would flock to university campuses to hire new graduate scientists, now annual conventions and personal recommendations have become the main source of employment. Recruitment meetings are usually demeaning and, by the very form in which they are run, produce hostile and aggressive attitudes among candidates competing for very few positions.

Several things are beginning to become evident from the overproduction or under-utilization of highly trained academics. No matter how sophisticated a man's education is, or how specialized his training, if he cannot find a place in the society which adequately rewards him for his educational achievement and continues to support his intellectual development, he suffers a loss of self-esteem and begins to see the system as an alien and pernicious force. We are in the era of industrialized science where an interlocking relationship exists between the practice of science and the economy. Academic science, characteristic of university financed research, is on the decline and with it scientific autonomy is declining as well.[18]

Bibliography

1. The major thrust of Marx's analysis of alienation appears in his *Economic and Philosophic Manuscripts of 1844.* For an excellent introduction by Dirk Struik see the edition by International Pub., New York: 1969.
2. Bentley Glass discusses the "growing isolation in which much scientific research is carried on," in "Information Crisis in Biology," *Bulletin of the Atomic Scientists,* Vol. 18 (Oct. 1962), pp. 6–12. The proliferation of scientific journals was being discussed as early as 1960. "How Many More New Journals?" *Nature,* Vol. 186 (April 2, 1960), pp. 18–19.
3. Special tutorial programs have been established to broaden the specialist's range of competence and provide him with a more comprehensive view of science. "What Do Other Specialists Do," *Physics Today,* Vol. 19 (March, 1966) pp. 43–46.
4. Unlike the situation in assembly-line industries where management wants workers to perform highly specialized tasks, industrial employers have expressed dissatisfaction with the narrowly trained breed of scientists who enter the job market from academia. See, for example, Arnold A. Strassenberg, "Supply and Demand for Physicists," *Physics Today,* Vol. 23 (April, 1970), pp. 23–28.
5. L.M. Andrews and Marvin Karlins, *Requiem for Democracy* (New York: Holt, Rinehart and Winston, 1971), p. 71.

6. J. Ben-David in *The Scientist's Role in Society* (Englewood Cliffs, N.J.: Prentice Hall, 1971) points out that the reason many scientists have experienced a feeling of loss of purpose and satisfaction in their work is their easy material success conjoined with little or no contribution to science and society.

7. This effect of specialization is discussed in a sociological study of engineers but applies equally well to scientists. "The increased specialization in technology has made it possible to escape social responsibilities by passing the buck—each specialist denies responsibility for the total product . . ." *Science,* Vol. 172 (June 11, 1971), p. 1103.

8. *The Structure of Scientific Revolutions* (Chicago: Univ. of Chicago Press, 1962), pp. 136–37.

9. Some attempts are being made to rethink the science curriculum. Harvard Project Physics, a curriculum development program directed at high school teachers, is designed to relate the teaching of physics to the history and philosophy of science. In general, however, any broadening of the science curriculum to integrate historical material has remained in its experimental stage. See, William H. Koch, "An Age of Change," *Physics Today,* Vol. 23 (Jan. 1970), p. 28. For an example of how physics can be taught through a historical approach by recreating classical experiments see, Samuel Devons and Lillian Hartman, "A History of Physics Laboratory," *Physics Today,* Vol. 23 (Feb. 1970), pp. 44–50.

10. Marx, *Economic and Philosophic Manuscripts,* p. 111.

11. A recent move against over-classifying and in the direction of giving society members more control over the type of classified research undertaken by the scientific community is witnessed by the resolution adopted by the National Academy of Sciences (NAS) that "gives the membership at large a means of participating in academy decisions to accept or reject contracts for classified projects." See, "NAS: Academy Votes NAC Changes, New Formula on Classified Research," *Science,* Vol. 176 (May 5, 1972), p. 499.

12. Niels Bohr, "The Quantum Postulate and the Recent Developments of Quantum Theory," *Nature,* Vol. 121 (April 14, 1928), pp. 580–590.

13. "What we cannot speak about we must pass over in silence." Ludwig Wittgenstein, *Tractatus Logico-Philosophicus.* Trans. by D.F. Pears and B.F. McGuinness (London: Routledge and Kegan Paul, 1966), p. 151.

14. Max Born, *The Born-Einstein Letters* (New York: Walker, 1971), pp. 144–147.

15. Marx, *Economic and Philosophic Manuscripts,* p. 108.

16. While projections for science employment vary, most studies agree that in the next two or three decades there will be an oversupply of highly trained scientific personnel. In one study a new physicist has only about one chance in ten of finding employment in his field. (Wayne R. Gruner, "Why There Is a Job Shortage," *Physics Today,* Vol. 23 (June 1970), pp. 21–26.) Another author concludes that "more physicists are dissatisfied with their jobs now, and even those jobs were more difficult to obtain than before." (Strassenberg, *Supply and Demand*).

17. For a report on the mood toward unionization in scientific societies see, "Professional Societies: Identity Crisis Threatens on Bread and Butter Issues," *Science,* Vol. 176 (May 19, 1972), pp. 777–779.

18. An illuminating discussion of the differences between academic and industrialized science is found in Jerome R. Ravetz, "Social Problems of Industrialized Science," *Scientific Knowledge and Its Social Problems* (Oxford: Clarendon Press, 1971). From the period 1953 to 1971 the government has taken an increasingly greater burden of the expenses of basic research. In 1953 it paid 45 percent of the cost for basic research in colleges and universities and in 1971 it covered 63 percent of the cost. For further data along with an argument that universities should have independent support funds for research, see Raymond J. Woodrow, "Government-University Financial Arrangements for Research," *Science,* Vol. 176 (May 26, 1972), pp. 885–889.

D. N. MICHAEL

Science, Scientists, and Politics

D.N. Michael is a social scientist and an author. He is associated with the Institute for Policy Studies in Washington, D.C. This essay addresses itself to the problem of formulating an adequate science policy. Present methods for formulating policy are found to be seriously flawed, in that the interests of the public are inadequately represented. Presently decisions about science policy at the federal level are seen as arising largely from groups within the scientific and political establishment; groups that are within themselves splintered and distrustful of each other.

Anthropologists and historians tell us that a crucial juncture in the life of a culture occurs when the assurance that it has gained from an unchallenged world view of values, goals, and logic confronts the unchallenged world view of another culture. It is not easy for men to change their view of the world, for it is part of their view of themselves. The challenge of other values threatens all that has given them comfort and support. It takes strong men and felicitous circumstances for a society to ride out the storm of contact with another culture and learn and grow anew.

It is by no means certain that this will happen. Some people are shattered by new experiences; so are some cultures. As segments of society splinter and converge, new institutions and new modes of thinking are generated. Some societies blossom in their revised form; others die.

Today we are faced with such a cultural crisis. The problems of making suitable policies for scientific work in the government arise chiefly from a profound cultural conflict. This conflict is the three-way confrontation among the scientific community, the nonscientific political governmental community, and the general public.

What is meant here by an adequate policy for federal science must be made clear at the outset. Such a policy would reconcile the needs of science and technology with the needs of the rest of society. Policy now springs from resolving disputes for priority among various projects. It is made in many places, from the Pentagon to the Department of Agriculture, as well as in those offices assigned part of the policy-making task. But nowhere do the social implications of science have a basic part in the formulation of policy.

Today, science and technology are not neutral. Not only does their development require vast social and human resources, but they are pursued because their powers for enhancing or degrading humanity are recognized. This nonneutrality demands an explicit relating of science and technology to the needs and proc-

"Science, Scientists, and Politics" by Donald N. Michael, 1963. Reprinted, with permission, from "Science, Scientists, and Politics," an Occasional Paper published by The Center for the Study of Democratic Institutions in Santa Barbara, California.

esses of society. This relationship should be the foundation of federal science policy.

The one concensus among the three cultures—the scientific community, the nonscientific political community, and the public—is that the task of government is to serve the general public. There is no such agreement about the relationship of science to government and to the general public. There is no set of values mutually subscribed to by the three cultures that defines the proper purposes of science and technology and thereby the appropriate restraints and supports needed to fulfill those purposes. Nor is it clear that such a set of values can be deliberately produced. Values do not derive solely from rational considerations. They are historical products of emotion and plain accident as much as, or more than, reason. This is one weakness in the thesis that the scientific method by itself can solve society's problems.

Within each of the three cultures are men and institutions with different viewpoints and different goals. These dissimilarities are crucial. Some of them derive largely from training; some are induced by the preconceptions that each group has about the other two and about itself. Two of the three are contending for the power to insure that their particular values will prevail: the science community and the nonscience governmental community. The general public has essentially no power.

The *science community* is represented at its upper levels by two types of scientists. The "traditional" type considers government to be synonymous with mediocrity and irrationality. These men feel that science must be left free to pursue its own ways. Their attitudes toward the rest of society are frequently ambivalent. They avoid involvement in social questions. Some of them perceive society as subject to, if not already operating along, logical lines. Others consider society as incorrigibly irrational and therefore unrelated to them. They are seldom asked to consider the social implications of their actions. By attending to their work, advising on the technical merits of this or that proposal, they can maintain the comfortable delusion that science can still be pursued without thought of the social consequences. Frequently they work for the university or for big industry, advancing the favorite programs of their employers.

Then there is the new breed of scientist around high Washington conference tables—the science entrepreneur, the "political" scientist. These men want to manage the bureaucracy to the extent necessary to make it behave the way they think it should. They have a sense of political technique, and they enjoy and seek power. Like the traditionalists, they feel that science is theirs, that no one else has the right to tamper with it. It is they who should decide which projects deserve emphasis. They believe a good dose of science would fix society fine, as C.P. Snow has so frequently tried to demonstrate. There are wise and modest men with social imagination in this subculture, but frequently the powerful members of this group are self-assured to the point of arrogance about their own abilities, about the over-riding rightness of scientific values and methods, and about the validity of their view of how society operates and what it needs.

The science entrepreneurs are supported by and in turn support big business, big publicity, big military, sometimes big academia and parts of big government.

They are both the captives and the kings of these powerful coalitions—kings for obvious reasons, captives because in reaping the benefits of affiliation they capitulate in some degree to the operating principles of these institutions. They have climbed to power through conservative hierarchies and tend to hold conservative values. The infusion of émigrés from the disciplined institutions of Europe seems, in general, not to have been a liberalizing influence. The more powerful the "political" scientist gets, the more omnipresent he is at major deliberations on science policy.

The *nonscientific community* in Congress and the bureaucracies regards itself as the bones, meat, and brains of government and society. They resent the "woolyheaded" scientist who may be trying to change their ways or implying that these ways are inadequate. They are not about to be displaced by a new attitude or a new kind of knowledge. Scientific expertise is respected, but the political and social naïveté that is supposed to accompany it is regarded with disdain. A general feeling exists among these "nonscientists" that science must be controlled. Usurpation of power is feared, partly because of a conviction that science somehow cannot be stopped.

These men consider society a nonrational environment. They see the political process as subtle and changing, responsive to many pressures of which science is only one, and by no means the most important. They view science as a means, not as an end. But they are confused about means and ends in general, as well as about the implications of science, and have no clear view of the proper role of scientists in formulating policy.

These two cultures between them decide on national science programs. They are in deep conflict within and between themselves. There are great political and ethical splinterings in the science community alone. The entrepreneurs claim to speak for science, but speak only for their faction. The traditionalists are fearful and envious of the "political" scientists, upon whom they must depend for their survival, especially if they hope for accomplishment in fields requiring expensive equipment or team research. Both groups are dissatisfied with the workings of government.

Given this clash of cultures, how can a valid basis be found for policy-making in federal science? We must discover a common ground from which science and technology can be intelligently directed. We must be able to evaluate the social consequences of scientific innovation. We need to plan our economics to assure the effective and humane introduction of modern technologies. We must equip government to meet new regulatory and managerial tasks. It is not clear that these responsibilities can be met by any traditional form of government; nor is it certain that democracy can be preserved in doing so. What is clear is that we cannot continue to bumble along.

Already we are in desperate trouble over nuclear weapons. We are about to be overwhelmed by that terrible blessing of medical technology, overpopulation. The social implications of biological and psycho-pharmacological engineering are already evident. Cybernation is causing serious problems. What is more, our environment is being changed in ways no cybernetical system can cope with indefinitely. It must respond to a tremendous and growing range of information at increasing speed and with increasing accuracy. Instability of the system is the inevitable result.

In spite of these menacing developments we remain unable to forecast the social consequences of technology. This is partly because of the limited vision of both the nonscientists and the scientists. The first group does not have sufficient knowledge of technology to sense the potentialities of new developments and therefore cannot predict their social impact, and they are too preoccupied with conventional assessments of political issues and impacts. The second group is aware of the technological possibilities but is not sufficiently sensitive to their social implications. Some of the scientists care only about the success of their favorite projects. Some apply to these problems a personal pseudo-sociology made useless by its arrogance or naïveté. And still others dodge responsibility by arguing that technology itself is neither good nor bad, that its virtues are determined by its uses.

Another reason why the social repercussions of science are difficult to forecast is that we have too little understanding of the social processes. This limitation has been fostered by the disinclination of the natural scientist and the government operator to stimulate work in the social sciences. The bureaucrat feels threatened by the possibility that formalized knowledge will replace "experience" and "political knowhow." Furthermore, the social sciences might demonstrate that the products of technology, or even science itself, need social control. This is an unhappy prospect for those scientists who are feeling for the first time the satisfactions of wielding power.

Since the consequences of scientific and technological developments are not fully predictable, it would seem impossible to establish priorities for individual projects on any sensible basis. Yet the forces of technological advance compel some kind of choice. Creative talent is a scarce resource, and the availability of money is a political, if not a real, limitation. "Political" scientists push their preferences vigorously, and the very existence of large programs influences selections in the absence of better criteria. Priority decisions today depend on political and economic pressures, personalities, and public relations.

The public relations juggernaut, in particular, imposes a crippling distortion on science and on those who would make scientific policy. From the laboratory to the launching pad, science and technology are harried by promises about "product superiority" and the glamour of "breakthroughs." Commitments are quickly publicized and then science is pressed to maintain the "reality" of the commitments. The natural failures of science and the natural limits of accomplishment are covered by an ever-deepening layer of misrepresentation, deviousness, and downright lies. So pervasive becomes the aura of untruth that it is hard for anyone, from the man in the laboratory to the public, to know where reality lies.

A cliché of our political folklore is that somehow the *public* will make everything right. In its wisdom it will judge between the contending power groups, evaluate technologies, establish a scale for priorities. But the public, the third culture, hardly knows what is happening. Understanding or judging the conflicts and compromises now occurring between science and government is far beyond its capacity. The public is caught between a publicity-induced fantasy world where science knows all the answers and a frustrating actuality which it does not realize is caused at least in part by the inadequate or incorrect use of science and technology. The frustrations are blamed on someone else: Russia, the government, perhaps the intellectuals, seldom on science. The public still believes in the mad scientist working on bombs, or in the humble scientist laboring over polio vaccine. The member of government, civil servant or politician, is perceived no more realistically.

Rather than becoming able to resolve the problems of science policy, the public is likely to become increasingly alienated both from government and from science. As with many other groups in the past that have met cultures somehow superior to their own, the public may withdraw from the challenge of "adjusting up" to the new priests and the new power. How, in fact, can the ordinary citizen adjust up to a computer-run society and classified questions of life and death?

One segment of the public will not surrender without protest. This is the group of articulate, concerned laymen who are not solely scientists, politicians, or civil servants and who worry about the arms race, overpopulation, the ascendancy of the "political" scientist, and the inadequacy of nonscientific bureaucracies. These people might be the moderators, the synthesists, for a new culture. They do not have the trained incapacities of those solely immersed in the two contending cultures, and they do have perspective that the general public lacks. But these very characteristics may deny them the opportunity. The day of the technical specialist grows ever brighter. The scientist will not freely yield his newly gained power, nor will the government worker relinquish his long-held dominion. Neither is likely to give ground to a nonspecialist who cannot build bombs or tread bureaucratic water, or otherwise play according to the rules of science and government.

The character of the coming generation of scientists is changing. The attributes attractive to laboratory directors interested in team-work are bringing a new personality into science. The old-guard traditionalists may be on the way out. Those who succeed will be those who are good at working with—or subverting—the nonscientific bureaucracy. Will these men be good scientists? This is not the important question. The real concern is for whom they will speak, and for what ends.

The problem in trying to resolve the ambitions of the two power cultures is that neither group has a clear view of what it wants in the way of policy for governmental science. As long as there is no community of values to guide judgment, basic policy decisions cannot be made, much less decisions on specific priorities for specific projects. Yet crises are arising on every hand. The evolution of a consensus cannot be awaited. If this society does not learn how to assimilate the changes that confront it, it will not survive.

DAVID RIESMAN

Leisure and Work in the Post-Industrial Society

David Riesman is Professor of Social Relations at Harvard University. He is the author of *The Lonely Crowd* (1950). Scientific and technological advances have, in our post-industrial society, created a situation in which arduous and boring work could be eliminated, but, as Riesman observes, this has not taken place—work has not been made more humanly satisfying for large segments of the population. Leisure time, in a society of abundance, also offers untold opportunities for improving the quality of life. Here, too, Riesman finds a depressing situation with the major emphasis in contemporary society being placed on conspicuous production and consumption.

Leisure—Society's Blotting Paper

It has become clear that post-industrial society no longer requires arduous and routinized work on the one hand, or, on the other hand, that kind of seemingly varied work, such as that of the salesman, in which the worker is compelled to exploit his own personality. Nevertheless, I have been arguing that Americans remain too unequivocally the children of industry, even when automation threatens to disinherit us, for us to be able to resort to leisure as a counterbalance for the deficiencies of work. Even so, leisure is coming to occupy for adults something of the position the school already occupies for youngsters, of being the institution which seems "available" to bear the brunt of all society's derelictions in other spheres. Thus, just as schools are asked to become quasi-parental, quasi-custodial, quasi-psychiatric, and quasi-everything else, filling in for tasks other institutions leave undone or badly done, with the result that the schools often cannot do their job of education adequately; so leisure is now being required to take up the energies left untapped everywhere else in our social order, with the result that it often fails in its original task of recreation for most of us most of the time and of creativity for some of us some of the time. The hopes I had put on leisure (in *The Lonely Crowd*) reflect, I suppose, my despair about the possibility of making work in modern society more meaningful and more demanding for the mass of men—a need which has come upon us so rapidly that the taste of abundance we have had in the past now threatens to turn into a glut.

My despair on this score, I must add, was not greatly alleviated by the feeling in the group of union leaders . . . that it was impossible either to get unions or management in the least interested in making work more humanly satisfying. I hoped the union leaders might cooperate with management in, so to speak, turning the engineers around, and forcing them to design men back into their machines rather than out of them. In this connection, I recall talking with aircraft engineers who were irritated with the "human factor," and eager to put a machine wherever a man might go wrong, rather than to design equipment that maximized the still enormous resourcefulness of the human mind. I recall the highway engineers who designed thruways that would look good to other engineers or to engineering-minded Americans—until the death toll made them realize that boredom could be a greater danger to man than speed and obstacles. And I thought of the subdividers who bulldozed down all trees to make it easier to build a road or a suburb, with no authorities around to forbid such wanton simplification of their own task along with such destruction of history and life. As the discussion with the union officials continued, it became clearer to me that the workers themselves were too much of this same school of engineering thought really to believe in the reorganization of industry. The kind of utopia of meaningful work pictured in Percival and Paul Goodman's book *Communitas* made no sense to them.

In this perspective, the rebellion of workers against modern industry is usually mere rebellion, mere goofing off. Many are quite prepared to go on wildcat strikes (Daniel Bell notes that in 1954–55 there were forty such in just one Westinghouse plant in East Pittsburgh); they are quite prepared to deceive the time-study man and to catch forty winks on the night shift, and otherwise to sabotage full production while still "making out" in terms of the group's norms— being in this like students who might cheat on exams or cut classes but could not conceive of reorganization of the curriculum or of asking for heavier assign- ments. The great victory of modern industry is that even its victims, the bored workers, cannot imagine any other way of organizing work, now that the tradition of the early nineteenth century Luddites, who smashed machines, has disappeared with the general acceptance of progress. We must thus think of restriction of output and other sabotage of production as mere symptoms.

Furthermore, the resentment which manifests itself in these symptoms helps engender a vicious circle, since it confirms the opinion of managment that workers must be disciplined by bringing them together in great factories and subjecting them to the relentless pressure of assembly lines—as against the possibility, for instance, that work could be decentralized so that workers would not have to commute long distances and could proceed more at their own pace and place. In the high-wage industries given over to "conspicuous production," management has the resources to be concerned with the amenities of work—the group harmony, the decor, the cafeteria and other ancillary services—and to make provision for the worker's leisure, such as bowling teams, golf courses, and adult education courses too; in fact, a whole range of extracurricular pleasures and benefits. Sometimes these benefits include profit-sharing, but they are much less likely to include decision-sharing, for of course managers object less to giving away money, especially money that would otherwise go to stockholders or to the government in taxes, than to giving away power and prestige and freedom of action to workers whose unionized demands reflect merely their discontent and scarcely at all their desires for reconstruction.

It is obvious in addition that managers are not free to reorganize their plants in order to provide their workers with a more satisfying work environment, if this might risk higher costs, unless their competitors are prepared to go along. Yet competition is not the whole story, for the situation is hardly better and is often worse in nationalized industries in Great Britain and Western Europe generally, while the situation of industrial workers in the Soviet Union today reminds one of the worst excesses of the Victorian era and the earlier days of the Industrial Revolution in the West. Managers of whatever ideological stripe seek to measure themselves against a single, unidimensional standard by which they can judge performance and thus are drawn to simplified work routines and an unremitting drive for maximum output. To open the possible consideration of factories as making not only things but also men, and as providing not only comfort and pay but also challenge and education, this would itself be a challenge to the way we have assimilated technology for the last three hundred years; and it would compel us to search for more Gestaltist and amorphous standards, in which we were no longer so clear as to what is process and what is product. There have, to be sure, been paternalistic employers (such as the Lowell mills in the 1840's or the Pullman plant a half-century ago) concerned with the education and uplift of their operatives—often to the eventual resentment and unionization of the latter (who felt it was enough to have to work for the bosses without imitating their preferred inhibitions). But these were efforts to compensate outside the plant for the dehumanization regarded as inevitable within. What I am asking for now is hardly less than reorganizing work itself so that man can live humanely on as well as off the job.

Strenuous Work

The work of the managers themselves, of course, striving to get out production in the face of technical and human obstacles, is seldom boring, although if the product itself is socially valueless, a point may be reached where work upon it, despite technical challenges, is felt as stultifying. Indeed, one could argue that the great disparities of privilege today are in the realm of the nature of work rather than in the nature of compensation: it has proved easier partially to equalize the latter through high-wage and tax policies than to begin at all on the former, which would require still greater readjustments. In that brilliant precursor of much contemporary science-fiction, Aldous Huxley's *Brave New World,* the lower cadres are given over to fairly undiluted hedonism while serious work and thought are reserved for the ruling "Alphas." Likewise, a recent science-fiction story once more illuminates the issue (it is my impression that science-fiction is almost the only genuinely subversive new literature in wide circulation today): this is a story by Frederick Pohl called "The Midas Plague" which pictures a society in which the upper classes are privileged by being allowed to spend less time and zeal in enforced consumption; they are permitted to live in smaller houses and to keep busy fewer robots in performing services for them. Their ration points—rations to extend rather than to limit consumption—are fewer; their cars are smaller; the things and gadgets that surround them are less oppressive. Best of all, they are allowed to work at work rather than having to spend four or five days a week simply as voracious consumers. That is, as one rises in the status system by excelling at consumership, one is allowed a larger and larger scope for what Veblen called the instinct of workmanship.

As already indicated, the world presented in "The Midas Plague," as in so much science-fiction, is all too little a fiction. For, if we except a number of farmers and skilled workers, such as tool and die makers, it is the professional and executive groups who at present have the most demanding and interesting work and for whom, at least until retirement, leisure is least a time to kill. The study by Nancy Morse and Robert Weiss referred to earlier indicates that on the whole these groups find most satisfaction in their work. A survey by *Fortune* last year showed that top executives, despite giving the appearance of being relaxed and taking it easy as our mores demand, work an average of sixty hours a week or more. In many other fields, the leisure revolution has increased the demands on those who service the leisure of others or who have charge of keeping the economy and the society, or considerable segments of it, from falling apart. High civil servants and diplomats probably work as hard or harder than ever— indeed it is not easy today to imagine writers like Hawthorne or Trollope holding civil service sinecures as a way of supporting themselves as novelists. Many priests and ministers, with expanding parishes and congregations, and with more expected of them in the way of ancillary services, find themselves as busy as any top executive. The same is true of a good many teachers and professors who are presumably training others to spend their leisure wisely! Physicians are notorious for their coronaries and their lack of care for their own health and comfort: as the public has more and more time to spend with doctors (often a kind of window shopping on themselves) and as there are fewer doctors in proportion to wealth and population, the medical men are forced to work seventy hours a week to pay for their monopoly position, their glamour, their high incomes, and their prestige. (The doctors at least have aides and antibiotics to help them out, but teachers and other ill-paid service workers have no similar labor-saving devices.) All in all, as I have suggested, those who are privileged in being able to choose their own work are becoming increasingly underprivileged with respect to leisure and perhaps also with respect to the pace at which, in the face of the waiting customers, they must respond to the demands upon them. A polarization is occurring between the toiling classes and the leisure masses.

In our egalitarian society, however, it would be surprising if the attitude of the masses did not influence the classes (there are of course also influences running the other way). As I remarked at the outset, I have the impression that a general decline is occurring in the zest for work, a decline which is affecting even those professional and intellectual groups whose complaint to their wives that they are overworked has often in the past been a way of concealing the fact that their work interested them rather more than did their wives. To return to the case of the doctors, for example, there is some slight evidence that application lists to medical school are no longer so full, a decline which is attributed to the belief among young people that medical education is too arduous and takes too long before one is stabilized on a plateau of suburban life and domesticity. Similar tendencies would appear to be affecting those already in medical school. Howard S. Becker and Blanche Geer report (from the study of medical education at the

University of Kansas being carried out under the direction of Professor Everett C. Hughes) that the teaching faculty complains that the students are no longer as interested in the more theoretical or scientific aspects of medicine: three-quarters of them are married and, instead of sitting around waiting for night duty or talking about their work, they are eager to go home, help the wife get dinner, and relax with television.

Likewise, there is evidence that young men in the big law firms, although they still work harder than most of their clients, do not glory in putting in night work and weekend hours as they once did. And several architects have told me that similar changes are showing up even in this field, which is famous for the enthusiasm of its devotees and the zest for work built up during *charettes* at architectural school. (Possibly, this may reflect in part the loss of the enthusiasm of the crusade on behalf of "modern" and the routinization of what had once been an esoteric creed.)

If such tendencies are showing up in the professions to which, in the past, men have been most devoted, it is not surprising that they should also be appearing in large-scale business enterprise. Though top executives may work as hard as ever— in part perhaps because, being trained in an earlier day, they can hardly help doing so—their subordinates are somewhat less work-minded. The recruiters who visit college campuses in order to sign up promising seniors or graduate students for large corporations have frequently noted that the students appear at least as interested in the fringe benefits as in the nature of the work itself; I would myself interpret this to signify that they have given up the notion that the work itself can be exciting and have an outlook which is not so very different from that of the typical labor union member: they want and expect more but not so very much more than the latter. Certainly, if fiction is any clue (of course, it is at best an unreliable clue) to prevailing attitudes, the current crop of business novels is revealing, for its indicates a marked change from an earlier era of energetic if ruthless tycoons. In *The Man in the Grey Flannel Suit,* for instance, the hero, Tom Rath, chooses the quiet suburban life and domesticity over the chance for large stakes and large decisions, but possible ulcers, in a big broad-casting company. He does so after discovering that his boss, powerful and dynamic, is estranged from everything in life that matters: from his wife, his daughter, and himself—his work is only an escape. Likewise, in *Executive Suite,* there is an analogous picture of the old tycoon who is wedded to his work and isolated by it, contrasted with the young hero whose work is at once not so strenuous and more playful and "creative."

The movement to the suburbs is of course a factor in these developments, especially now that young men move to the suburbs not only for the sake of their wives and young children and the latter's schooling but also for their own sake. It is hard, for example, for a scientific laboratory to maintain a nighttime climate of intense intellectual enthusiasm when its professional cadres are scattered over many suburbs and when the five-day week has become increasingly standard throughout American life (outside of a few universities which cling to the older five and a half day pattern). The sport-shirted relaxed suburban culture presents a standing "reproach" to the professional man who works at night and

Saturdays instead of mowing the lawn, helping the Little League baseball team, and joining in neighborly low-pressure sociability. The suburbs continue the pattern of the fraternity house in making it hard for an individual to be a ratebuster or an isolate.

It is difficult to form a just estimate of the extent and scope of these changes. It is not new for the older generation to bewail the indolence of the young, and there is a tendency for the latter to maintain much of the older ethic screened by a new semantics and an altered ideology. Moreover, Americans in earlier periods were not uniformly work-minded. In Horace Greeley's account of his famous trip West in 1859 (which ended in his interviewing Brigham Young), he commented with disgust on the many squatters on Kansas homesteads who, in contrast to the industrious Mormons, sat around improvidently, building decent shelter neither for themselves nor for their stock (they sound a bit like Erskine Caldwell types). Similarly, the correspondence of railroad managers in the last century (and railroad managers were perhaps the most professional managerial groups as they were in charge of the largest enterprises) is full of complaints about the lack of labor discipline; this is one reason that the Chinese were brought in to work on the transcontinental roads. There were, it is evident, many backsliders in the earlier era from the all-pervading gospel of work, and the frontier, like many city slums, harbored a number of drifters. Today, in contrast, the gospel of work is far less tenacious and overbearing, but at the same time the labor force as a whole is post-industrial in the sense of having lost much of its pre-industrial resistance to the clock and to factory discipline generally.

Strenuous Leisure

So far, I have largely been discussing the uneven distribution of leisure in terms of differential attitudes towards work in different occupational groups. In comparison with the achievements of our occupational sociology, however, we have little comparable information concerning the sociology of leisure. For instance, we have very few inventories of how leisure is actually spent (apart from fairly complete information concerning exposure to the mass media). Pitirim Sorokin before World War II and more recently Albert J. Reiss, Jr. have tried to get people to keep diaries which would include accounts of their day-by-day use of leisure time; but these suffer from faulty memory and stereotyping (people often say, "one day is just like another," and report accordingly) as well as from omissions of fights and other improper activities. A more systematic study than most, by Alfred Clarke, found that radio and TV listening were the top two activities for both upper and lower prestige groups, followed by studying in the upper group and do-it-yourself activities in the lower. The latter spend much more time just driving around, as well as polishing the car; they also spend much more time in taverns. Only in the upper group do people go out to parties, as against simply dropping in on a neighbor to look at TV or chat in the kitchen; and going to meetings is also largely confined to the upper group. In both groups, commercial recreation outside the home, such as going to the movies, plays little part. This and other, more impressionistic studies point to the conclusion that the busier people, the professional and executives and better-educated groups generally, also lead a more active life in their time away from work; as the saying goes, they work hard and play hard. In Reiss's study, for example, there turned up a surgeon at a leading hospital who went to mass every morning, then to the hospital, then to attend to his private practice; he belonged to about every community organization, and he and his wife entertained three or four nights a week. Contrastingly, at the other end of the social scale, the unemployed as we know from several studies have in a psychological sense no leisure time at all; they, and the underprivileged generally, do not belong to voluntary associations (churches and unions are an occasional exception); they live what is often a shorter life on a slower timetable.

At the same time, as I have indicated above, it is among the less privileged groups relatively new to leisure and consumption that the zest for possessions retains something of its pristine energy. Consumership which is complex if not jaded among the better-educated strata seems to be relatively unequivocal among those recently released from poverty and constriction of choice. . . . With very little hope of making work more meaningful, these people look to their leisure time and consumership for the satisfactions and pride previously denied them by the social order.

I am suggesting here that millions of Americans, coming suddenly upon the inheritance of abundance, are able like other nouveau riche people in the past to coast upon the goals set out for them by their social and economic pace-setters. "Coast" is perhaps not the right word for so energetic a movement, one which continues to power the economy, as millions are moved out of dire poverty and subsistence into the strata which have some discretionary spending power; while in better educated strata the absence of goals for leisure and consumption is beginning, or so I would contend, to make itself felt. In these latter groups, it is no longer so easy to regard progress simply in terms of "more": more money, more free time, more things. There is a search for something more real as the basis for life, a search reflected in the vogue of psychoanalysis, of self-help books (and, in a few circles, of theology), of the growth of adult education courses which are nonvocational, and in the more serious nonfiction reading which is reflected in many of the new series of paperbound books. Such Americans are not satisfied simply to attain material comfort far beyond what their parents possessed or beyond what is obtained in most parts of the globe. In fact, the younger generation of reasonably well-off and well-educated Americans do not seem to me drivingly or basically materialistic; they have little ferocious desire for things for their own sake, let alone money or land for its own sake. At most it could be said that such Americans resent being deprived of those things they are supposed to have; consequently, they remain susceptible to advertising which tempts them with the halo of experience or associations surrounding goods— although not with the goods themselves as sheer objects. Hence in these strata there is a tendency for people, once accustomed to upper-middle-class norms, to lose eagerness for bounteous spending on consumer goods. Moreover, such Americans tend more and more to secure their children's future, not by large capital acquisitions and inheritances, but by giving them a good education and

the motives for achievement that go with it; they will try to pass on their values
as an insurance of continuing middle-class position, rather than their possessions
and their specific place in the occupational scheme. It is in such relatively
sophisticated Americans that we can see foreshadowed a decline of interest in
material goods that may be a long time appearing in the working class and lower
white-collar groups.

Indeed, the amenities which such educated people desire, once their own
families are well provided for, are not those which can be bought by individuals
acting in isolation from each other. They are rather such social goods as pleasant
cities and sprawl-free countrysides; adequate public services, including transport;
educational and cultural facilities which stimulate all ages and stages; freedom
from crowding in the sites of leisure; and in general, wise and magnanimous use of
the surplus which individuals at this level no longer need. But it is just at this
point that the paucity of our individual goals, when amplified at the general
social level, creates the most terrifying problems.

Abundance for What?

Even the most confident economists cannot adequately picture a society which
could readily stow away the goods likely to descend upon us in the next fifteen
years (assuming only a modest rise in annual productivity), with any really
sizable drop in defense expenditures. People who are forced by the recession or
by fear of their neighbors' envy or by their own misgivings to postpone for a
year the purchase of a new car may discover that a new car every three years
instead of every two is quite satisfactory. And once they have two cars, a swim-
ming pool and a boat, and summer and winter vacations, what then?

Increasingly, as we all know, the motivation researchers are being pressed to
answer these questions, and to discover what the public does not yet know that
it "wants." Just as we are lowering our water table by ever-deeper artesian wells
and in general digging ever deeper for other treasures of the earth, so we are
sinking deeper and deeper wells into people in the hope of coming upon "motives"
which can power the economy or some particular sector of it. I am suggesting
that such digging, such forcing emotions to externalize themselves, cannot
continue much longer without man running dry.

Even now, some of the surplus whose existence presents us with such questions
is being absorbed in the very process of its creation but by what I have termed
the "conspicuous production" of our big corporations, acting as junior partners
of the welfare state and absorbing all sorts of ancillary services for their own
members and their own communities.

Defense expenditures loom so large in our political as well as economic horizon
because they do offer an easy and seemingly feasible way out by creating goods
which do not themselves create other goods. (They are "multipliers" only in a
Keynesian sense.) But of course the international consequences as well as the
long-range domestic ones point the way only to lunacy and the alternatives of
destruction or the garrison state. Indeed in a recent article, "Economic Implications of Disarmament," Gerard Colm argues that it would be difficult to deploy

for public services our rising productivity even without reducing defense expenditures. He sees education as potentially absorbing much the largest part of the surplus (education must be seen even now as the greatest leisure time-killer we have, keeping out of the labor force an increasingly large portion of the young). And Colm presents figures for highway and other transport, along with other public works, hospitals, and water conservation—yet these altogether hardly make up in ten years what we spend in one year for our armed forces. I would contend that expenditures which serve no real social imperative, other than propping up the economy or subduing the sibling rivalry of the armed services, will eventually produce wasteful byproducts to slow that economy down in a tangle of vested inefficiencies, excessively conspicuous production, lowered work morale, and lack of purpose and genuine inventiveness. The word "to soldier" means "to loaf" and conscription gives training in soldiering to a large proportion of the future work-force (despite islands of ascetism in the Strategic Air Command or the air-borne "brushfire" infantry). For a time, men will go on producing because they have got the habit, but the habit is not contagious. Men will scarcely go on producing as mere items in a multiplier effect or conscripts in an endless Cold War, nor will they greatly extend themselves to earn more money which they are increasingly bored with spending. To be sure, many workers have little objection to getting paid without "earning" it by traditional standards of effortfulness. And while those standards are usually irrelevant in a society of advanced technology and high expenditures on research and development, there are certainly many parts of the economy, notably in the service trades, whose gross inefficiency we only conceal from ourselves by contrasting America with pre-industrial societies or with those possessing far less adequate resources of men and machines—if we compare ourselves with the West Germans, for instance, or with the Canadians, the advance in our economy since 1946, great as it is in absolute terms, is unimpressive. The pockets of efficiency in our society are visible and made more so by the myth that we are efficient; hence, the evidence of disintegration and incompetence that is all around us strikes us as temporary or aberrant.

The Dislocation of Desire

Correspondingly, some of our desires have been made highly visible by advertising and market research and lead to equally visible results such as good cars and, intermittently, good roads to drive them over. But other desires, which require cooperation to be effective, are often lamely organized and all but invisible. Thus, while some of us have a missionary zeal for learning, which we regard as the basis of later leisure as well as later employment, we have not been helped even by the push of sputnik to get a bill for school construction past the same Congress which eagerly voted Federal money for highways (in part, no doubt, because the annual maintenance of schools falls upon a local tax base which grows constantly more inadequate while the maintenance of highways can be more easily financed from gasoline and registration taxes). Other services, not so clearly "a good thing" as secondary and university education, are even more

lacking in organized institutional forms which would permit the channeling of
our surplus in ways which would improve the quality and texture of daily life.
For example, even the great demand for scenic beauty (anemically reflected in
the new highways) cannot make itself politically felt in preserving the countryside against roadside slums and metropolitan expansion, while men of wealth are
missing who could buy up whole counties and give them to the nation as a
national park. We see one consequence on summer weekends when millions pour
onto the roads and breathe each other's fumes and crowd each other's resorts.
And we see too that leisure is cut down by the time taken to get to and from
work—commuting time increased by the desire to live in the suburbs in order to
enjoy leisure! As our resources dwindle in comparison with population and as
individual abundance creates social blight, we will increasingly find little solace
in leisure without privacy. It is extraordinary how little we have anticipated the
problems of the bountiful future, other than to fall back on remedies which did
not work in the less bountiful past, such as individualism, thrift, hard work, and
enterprise on the one side, or harmony, togetherness, and friendliness on the
other. Meanwhile, we stave off the fear of satiation in part by scanning the technological horizon for new goods which we will all learn to want, in part by the
delaying tactic of a semi-planned recession, and, as already indicated, in part by
the endless race of armaments.

That race has its cultural as well as Keynesian dynamic: as poll data show, a
majority or large plurality of Americans expect war, though perhaps in a rather
abstract way—war is one of those extrapolations from the past; like technological
progress, we find it hard to resist. And, on the one hand, the threat of war is one
factor in discouraging long-term plans, while, on the other hand, the continuation
of the Cold War provides a sort of alternative to planning. Thus, there tends to be
a state of suspended animation in the discussion concerning the quality of life a
rich society should strive for; social inventiveness tends to be channeled into the
defense of past gains rather than into ideas for a better proportionality between
leisure and work. Like soldiers off duty, "as you were," we subsist in default of
more humane hopes.

But I should add that no society has even been in the same position as ours, of
coming close to fulfilling the age-old dream of freedom from want, the dream of
plenty. And I want to repeat that millions of Americans, perhaps still the great
majority, find sufficient vitality in pursuit of that dream: the trip to the end of
the consumer rainbow retains its magic for the previously deprived. It is only the
minority where, looking further ahead, we can see already the signs of a generation, prepared for Paradise Lost, which does not know what to do with Paradise
Found. Regrettably, it will not be given a long time to come to a decision. For,
by concentrating all energies on preserving freedom from want and pressing
for consumer satiation at ever more opulent levels, we jeopardize this achievement in a world where there are many more poor nations than rich ones and in
which there are many more desires, even among ourselves, for things other than
abundance.

HERBERT MARCUSE

Liberation from the Affluent Society

Herbert Marcuse, an influential philosopher, and formerly Professor of Philosophy at Brandeis University in Massachusetts and University of California at San Diego, has written a number of important books including *Reason and Revolution* (1941), *Eros and Civilization* (1955), and *One-Dimensional Man* (1964). Technological advances in industrial societies have, in Marcuse's analysis, brought about a new enslavement of man; an enslavement that, because of its technological base, is effective to an unprecedented degree. Marcuse calls for liberation, and sees this liberation as possible only through a qualitative change in society's organization. This qualitative change would mean a change from the present capitalistic organization of society to a socialist organization defined in its most Utopian terms.

I am very happy to see so many flowers here and that is why I want to remind you that flowers, by themselves, have no power whatsoever, other than the power of men and women who protect them and take care of them against aggression and destruction.

As a hopeless philosopher for whom philosophy has become inseparable from politics, I am afraid I have to give here today a rather philosophical speech, and I must ask your indulgence. We are dealing with the dialectics of liberation (actually a redundant phrase, because I believe that all dialectic is liberation) and not only liberation in an intellectual sense, but liberation involving the mind and the body, liberation involving entire human existence. Think of Plato: the liberation from the existence in the cave. Think of Hegel: liberation in the sense of progress and freedom on the historical scale. Think of Marx. Now in what sense is all dialectic liberation? It is liberation from the repressive, from a bad, a false system—be it an organic system, be it a social system, be it a mental or intellectual system: liberation by forces developing within such a system. That is a decisive point. And liberation by virtue of the contradiction generated by the system, precisely because it is a bad, a false system.

I am intentionally using here moral, philosophical terms, values: 'bad', 'false'. For without an objectively justifiable goal of a better, a free human existence, all liberation must remain meaningless—at best, progress in servitude. I believe that in Marx too socialism *ought* to be. This 'ought' belongs to the very essence of scientific socialism. It *ought* to be; it is, we may almost say, a biological, sociological and political necessity. It is a biological necessity in as much as a socialist society, according to Marx, would conform with the very *logos* of life, with the essential possibilities of a human existence, not only mentally, not only intellectually, but also organically.

"Liberation from the Affluent Society" by Herbert Marcuse as included in *To Free a Generation* edited by David Cooper. Copyright ©1968 by The Institute of Phenomenological Studies. Reprinted by permission of Deborah Rogers Ltd., London. David Cooper, ed., *To Free a Generation.* New York: The Macmillan Company, 1968, pp. 175–192.

Now as to today and our own situation. I think we are faced with a novel situation in history, because today we have to be liberated from a relatively well-functioning, rich, powerful society. I am speaking here about liberation from the affluent society, that is to say, the advanced industrial societies. The problem we are facing is the need for liberation not from a poor society, not from a disintegrating society, not even in most cases from a terroristic society, but from a society which develops to a great extent the material and even cultural needs of man—a society which, to use a slogan, delivers the goods to an ever larger part of the population. And that implies, we are facing liberation from a society where liberation is apparently without a mass basis. We know very well the social mechanisms of manipulation, indoctrination, repression which are responsible for this lack of a mass basis, for the integration of the majority of the oppositional forces into the established social system. But I must emphasize again that this is not merely an ideological integration; that it is not merely a social integration; that it takes place precisely on the strong and rich basis which enables the society to develop and satisfy material and cultural needs better than before.

But knowledge of the mechanisms of manipulation or repression, which go down into the very unconscious of man, is not the whole story. I believe that we (and I will use 'we' throughout my talk) have been too hesitant, that we have been too ashamed, understandably ashamed, to insist on the integral, radical features of a socialist society, its qualitative difference from all the established societies: the qualitative difference by virtue of which socialism is indeed the negation of the established systems, no matter how productive, no matter how powerful they are or they may appear. In other words—and this is one of the many points where I disagree with Paul Goodman—our fault was not that we have been too immodest, but that we have been too modest. We have, as it were, repressed a great deal of what we should have said and what we should have emphasized.

If today these integral features, these truly radical features which make a socialist society a definite negation of the existing societies, if this qualitative difference today appears as Utopian, as idealistic, as metaphysical, this is precisely the form in which these radical features must appear if they are really to be a definite negation of the established society: if socialism is indeed the rupture of history, the radical break, the leap into the realm of freedom—a total rupture.

Let us give one illustration of how this awareness, or half-awareness, of the need for such a total rupture was present in some of the great social struggles of our period. Walter Benjamin quotes reports that during the Paris Commune, in all corners of the city of Paris there were people shooting at the clocks on the towers of the churches, palaces and so on, thereby consciously or half-consciously expressing the need that somehow time has to be arrested; that at least the prevailing, the established time continuum has to be arrested, and that a new time has to begin—a very strong emphasis on the qualitative difference and on the totality of the rupture between the new society and the old.

In this sense, I should like to discuss here with you the repressed prerequisites of qualitative change. I say intentionally 'of qualitative change,' not 'of revolution,' because we know of too many revolutions through which the continuum of repression has been sustained, revolutions which have replaced one system of domination by another. We must become aware of the essentially new features which distinguish a free society as a definite negation of the established societies, and we must begin formulating these features, no matter how metaphysical, no matter how Utopian, I would even say no matter how ridiculous we may appear to the normal people in all camps, on the right as well as on the left.

What is the dialectic of liberation with which we here are concerned? It is the construction of a free society, a construction which depends in the first place on the prevalence of the vital need for abolishing the established systems of servitude; and secondly, and this is decisive, it depends on the vital commitment, the striving, conscious as well as sub- and unconscious, for the qualitatively different values of a free human existence. Without the emergence of such new needs and satisfactions, the needs and satisfactions of free men, all change in the social institutions, no matter how great, would only replace one system of servitude by another system of servitude. Nor can the emergence—and I should like to emphasize this—nor can the emergence of such new needs and satisfactions be envisaged as a mere by-product, the mere result, of changed social institutions. We have seen this, it is a fact of experience. The development of the new institutions must already be carried out and carried through by men with the new needs. That, by the way, is the basic idea underlying Marx's own concept of the proletariat as the historical agent of revolution. He saw the industrial proletariat as the historical agent of revolution, not only becuase it was the basic class in the material process of production, not only because it was at that time the majority of the population, but also because this class was 'free' from the repressive and aggressive competitive needs of capitalist society and therefore, at least potentially, the carrier of essentially new needs, goals and satisfactions.

We can formulate this dialectic of liberation also in a more brutal way, as a vicious circle. The transition from voluntary servitude (as it exists to a great extent in the affluent society) to freedom presupposes the abolition of the institutions and mechanism of repression. And the abolition of the institutions and mechanisms of repression already presupposes liberation from servitude, prevalence of the need for liberation. As to needs, I think we have to distinguish between the need for changing intolerable conditions of existence, and the need for changing the society as a whole. The two are by no means identical, they are by no means in harmony. *If* the need is for changing intolerable conditions of existence, with at least a reasonable chance that this can be achieved within the established society, with the growth and progress of the established society, then this is merely quantitative change. Qualitative change is a change of the very system as a whole.

I would like to point out that the distinction between quantitative and qualitative change is not identical with the distinction between reform and revolution. Quantitative change can mean and can lead to revolution. Only the conjunction, I suggest, of these two is revolution in the essential sense of the leap from prehistory into the history of man. In other words, the problem with which we are faced is the point where quantity can turn into quality, where the quantitative change in the conditions and institutions can become a qualitative change affecting all human existence.

Today the two potential factors of revolution which I have just mentioned are disjointed. The first is most prevalent in the underdeveloped countries, where quantitative change—that is to say, the creation of human living conditions—is in itself qualitative change, but is not yet freedom. The second potential factor of revolution, the prerequisites of liberation, are potentially there in the advanced industrial countries, but are contained and perverted by the capitalist organization of society.

I think we are faced with a situation in which this advanced capitalist society has reached a point where quantitative change can technically be turned into qualitative change, into authentic liberation. And it is precisely against this truly fatal possibility that the affluent society, advanced capitalism, is mobilized and organized on all fronts, at home as well as abroad.

Before I go on, let me give a brief definition of what I mean by an affluent society. A model, of course, is American society today, although even in the U.S. it is more a tendency, not yet entirely translated into reality. In the first place, it is a capitalist society. It seems to be necessary to remind ourselves of this because there are some people, even on the left, who believe that American society is no longer a class society. I can assure you that it is a class society. It is a capitalist society with a high concentration of economic and political power; with an enlarged and enlarging sector of automation and coordination of production, distribution and communication; with private ownership in the means of production, which, however, depends increasingly on ever more active and wide intervention by the government. It is a society in which, as I mentioned, the material as well as cultural needs of the underlying population are satisfied on a scale larger than ever before—but they are satisfied in line with the requirements and interests of the apparatus and of the powers which control the apparatus. And it is a society growing on the condition of accelerating waste, planned obsolescence and destruction, while the substratum of the population continues to live in poverty and misery.

I believe that these factors are internally interrelated, that they constitute the syndrome of late capitalism: namely, the apparently inseparable unity—inseparable for the system—of productivity and destruction, of satisfaction of needs and repression, of liberty within a system of servitude—that is to say, the subjugation of man to the apparatus, and the inseparable unity of rational and irrational. We can say that the rationality of the society lies in its very insanity, and that the insanity of the society is rational to the degree to which it is efficient, to the degree to which it delivers the goods.

Now the question we must raise is: why do we need liberation from such a society if it is capable—perhaps in the distant future, but apparently capable—of conquering poverty to a greater degree than ever before, of reducing the toil of labor and the time of labor, and of raising the standard of living? If the price for all goods delivered, the price for this comfortable servitude, for all these achievements, is exacted from people far away from the metropolis and far away from its affluence? If the affluent society itself hardly notices what it is doing, how it is spreading terror and enslavement, how it is fighting liberation in all corners of the globe?

We know the traditional weakness of emotional, moral and humanitarian arguments in the face of such technological achievement, in the face of the irrational rationality of such a power. These arguments do not seem to carry any weight against the brute facts—we might say brutal facts—of the society and its productivity. And yet, it is only the insistence on the real possibilities of a free society, which is blocked by the affluent society—it is only this insistence in practice as well as in theory, in demonstration as well as in discussion, which still stands in the way of the complete degradation of man to an object, or rather subject/object, of total administration. It is only this insistence which still stands in the way of the progressive brutalization and moronization of man. For—and I should like to emphasize this—the capitalist Welfare State is a Warfare State. It must have an Enemy, with a capital E, a total Enemy; because the perpetuation of servitude, the perpetuation of the miserable struggle for existence in the very face of the new possibilities of freedom, activates and intensifies in this society a primary aggressiveness to a degree, I think, hitherto unknown in history. And this primary aggressiveness must be mobilized in socially useful ways, lest it explode the system itself. Therefore the need for an Enemy, who must be there, and who must be created if he does not exist. Fortunately, I dare say, the Enemy does exist. But his image and his power must, in this society, be inflated beyond all proportions in order to be able to mobilize this aggressiveness of the affluent society in socially useful ways.

The result is a mutilated, crippled and frustrated human existence: a human existence that is violently defending its own servitude.

We can sum up the fatal situation with which we are confronted. Radical social change is objectively necessary, in the dual sense that it is the only chance to save the possibilities of human freedom and, furthermore, in the sense that the technical and material resources for the realization of freedom are available. But while this objective need is demonstrably there, the subjective need for such a change does not prevail. It does not prevail precisely among those parts of the population that are traditionally considered the agents of historical change. The subjective need is repressed, again on a dual ground: firstly, by virtue of the actual satisfaction of needs, and secondly, by a massive scientific manipulation and administration of needs—that is, by a systematic social control not only of the consciousness, but also of the unconscious of man. This control has been made possible by the very achievements of the greatest liberating sciences of our time, in psychology, mainly psychoanalysis and psychiatry. That they could become and have become at the same time powerful instruments of suppression, one of the most effective engines of suppression, is again one of the terrible aspects of the dialectic of liberation.

This divergence between the objective and the subjective need changes completely, I suggest, the basis, the prospects and the strategy of liberation. This situation presupposes the emergence of new needs, qualitatively different and even opposed to the prevailing aggressive and repressive needs: the emergence of a new type of man, with a vital, biological drive for liberation, and with a consciousness capable of breaking through the material as well as ideological veil of the affluent society. In other words, liberation seems to be predicated upon the opening and the activation of a depth dimension of human existence, this side of and underneath the traditional material base: not an idealistic dimension, over and above the material base, but a dimension even more material than the material base, a dimension underneath the material base. I will illustrate presently what I mean.

The emphasis on this new dimension does not mean replacing politics by psychology, but rather the other way around. It means finally taking account of the fact that society has invaded even the deepest roots of individual existence, even the unconscious of man. *We* must get at the roots of society in the individuals themselves, the individuals who, because of social engineering, constantly reproduce the continuum of repression even through the great revolution.

This change is, I suggest, not an ideological change. It is dictated by the actual development of an industrial society, which has introduced factors which our theory could formerly correctly neglect. It is dictated by the actual development of industrial society, by the tremendous growth of its material and technical productivity, which has surpassed and rendered obsolete the traditional goals and preconditions of liberation.

Here we are faced with the question: Is liberation from the affluent society identical with the transition from capitalism to socialism? The answer I suggest is: It is not identical, if socialism is defined merely as the planned development of the productive forces, and the rationalization of resources (although this remains a precondition for all liberation). It is identical with the transition from capitalism to socialism, if socialism is defined in its most Utopian terms: namely, among others, the abolition of labor, the termination of the struggle for existence—that is to say, life as an end in itself and no longer as a means to an end—and the liberation of human sensibility and sensitivity, not as a private factor, but as a force for transformation of human existence and of its environment. To give sensitivity and sensibility their own right is, I think, one of the basic goals of integral socialism. These are the qualitatively different features of a free society. They presuppose, as you may already have seen, a total transvaluation of values, a new anthropology. They presuppose a type of man who rejects the performance principles governing the established societies; a type of man who has rid himself of the aggressiveness and brutality that are inherent in the organization of established society, and in their hypocritical, puritan morality; a type of man who is biologically incapable of fighting wars and creating suffering; a type of man who has a good conscience of joy and pleasure, and who works, collectively and individually, for a social and natural environment in which such an existence becomes possible.

The dialectic of liberation, as turned from quantity into quality, thus involves, I repeat, a break in the continuum of repression which reaches into the depth dimension of the organism itself. Or, we may say that today qualitative change, liberation, involves organic, instinctual, biological changes at the same time as political and social changes.

The new needs and satisfactions have a very material basis, as I have indicated.

They are not thought out but are the logical derivation from the technical, material and intellectual possibilities of advanced, industrial society. They are inherent in, and the expression of, the productivity of advanced industrial society, which has long since made obsolete all kinds of inner-worldly asceticism, the entire work discipline on which Judaeo-Christian morality has been based.

Why is this society surpassing and negating this type of man, the traditional type of man, and the forms of his existence, as well as the morality to which it owes much of its origins and foundations? This new, unheard-of and not anticipated productivity allows the concept of a technology of liberation. Here I can only briefly indicate what I have in mind: such amazing and indeed apparently Utopian tendencies as the convergence of technique and art, the convergence of work and play, the convergence of the realm of necessity and the realm of freedom. How? No longer subjected to the dictates of capitalist profitability and of efficiency, no longer to the dictates of scarcity, which today are perpetuated by the capitalist organization of society; socially necessary labor, material production, would and could become (we see the tendency already) increasingly scientific. Technical experimentation, science and technology would and could become a play with the hitherto hidden—methodically hidden and blocked— potentialities of men and things, of society and nature.

This means one of the oldest dreams of all radical theory and practice. It means that the creative imagination, and not only the rationality of the performance principle, would become a productive force applied to the transformation of the social and natural universe. It would mean the emergence of a form of reality which is the work and the medium of the developing sensibility and sensitivity of man.

And now I throw in the terrible concept: it would mean an 'aesthetic' reality— society as a work of art. This is the most Utopian, the most radical possibility of liberation today.

What does this mean, in concrete terms? I said, we are not concerned here with private sensitivity and sensibility, but with sensitivity and sensibility, creative imagination and play, becoming forces of transformation. As such they would guide, for example, the total reconstruction of our cities and of the countryside; the restoration of nature after the elimination of the violence and destruction of capitalist industrialization; the creation of internal and external space for privacy, individual autonomy, tranquillity; the elimination of noise, of captive audiences, of enforced togetherness, of pollution, of ugliness. These are not—and I cannot emphasize this strongly enough— snobbish and romantic demands. Biologists today have emphasized that these are organic needs for the human organism, and that their arrest, their perversion and destruction by capitalist society, actually mutilates the human organism, not only in a figurative way but in a very real and literal sense.

I believe that it is only in such a universe that man can be truly free, and truly human relationships between free beings can be established. I believe that the idea of such a universe guided also Marx's concept of socialism, and that these aesthetic needs and goals must from the beginning be present in the reconstruction of society, and not only at the end or in the far future. Otherwise, the needs

and satisfactions which reproduce a repressive society would be carried over into
the new society. Repressive men would carry over their repression into the new
society.

Now, at this farthest point, the question is: how can we possibly envisage the
emergence of such qualitatively different needs and goals as organic, biological
needs and goals and not as superimposed values? How can we envisage the
emergence of these needs and satisfactions within and against the established
society—that is to say, prior to liberation? That was the dialectic with which I
started, that in a very definite sense we have to be free from in order to create
a free society.

Needless to say, the dissolution of the existing system is the precondition for
such qualitative change. And the more efficiently the repressive apparatus of
the affluent societies operates, the less likely is a gradual transition from servitude
to freedom. The fact that today we cannot identify any specific class or any
specific group as a revolutionary force, this fact is no excuse for not using any
and every possibility and method to arrest the engines of repression in the
individual. The diffusion of potential opposition among the entire underlying
population corresponds precisely to the total character of our advanced capitalist
society. The internal contradictions of the system are as grave as ever before and
likely to be aggravated by the violent expansion of capitalist imperialism. Not
only the most general contradictions between the tremendous social wealth on
the one hand, and the destructive, aggressive and wasteful use of this wealth on
the other; but far more concrete contradictions such as the necessity for the
system to automate, the continued reduction of the human base in physical
labor-power in the material reproduction of society and thereby the tendency
towards the draining of the sources of surplus profit. Finally, there is the threat
of technological unemployment which even the most affluent society may no
longer be capable of compensating by the creation of ever more parasitic and
unproductive labor: all these contradictions exist. In reaction to them suppression, manipulation and integration are likely to increase.

But fulfilment is there, the ground can and must be prepared. The mutilated
consciousness and the mutilated instincts must be broken. The sensitivity and the
awareness of the new transcending, antagonistic values—they are there. And they
are there, they are here, precisely among the still nonintegrated social groups and
among those who, by virtue of their privileged position, can pierce the ideological
and material veil of mass communication and indoctrination—namely, the
intelligentsia.

We all know the fatal prejudice, practically from the beginning, in the labor
movement against the intelligentsia as catalyst of historical change. It is time to
ask whether this prejudice against the intellectuals, and the inferiority complex
of the intellectuals resulting from it, was not an essential factor in the development of the capitalist as well as the socialist societies: in the development and
weakening of the opposition. The intellectuals usually went out to organize the
others, to organize in the communities. They certainly did not use the potenti-
ality they had to organize themselves, to organize among themselves not only on
a regional, not only on a national, but on an international level. That is, in my

view, today one of the most urgent tasks. Can we say that the intelligentsia is the agent of historical change? Can we say that the intelligentsia today is a revolutionary class? The answer I would give is: No, we cannot say that. But we can say, and I think we must say, that the intelligentsia has a decisive preparatory function, not more; and I suggest that this is plenty. By itself it is not and cannot be a revolutionary class, but it can become the catalyst, and it has a preparatory function—certainly not for the first time, that is in fact the way all revolution starts—but more, perhaps, today than ever before. Because—and for this too we have a very material and very concrete basis—it is from this group that the holders of decisive positions in the productive process will be recruited, in the future even more than hitherto. I refer to what we may call the increasingly scientific character of the material process of production, by virtue of which the role of the intelligentsia changes. It is the group from which the decisive holders of decisive positions will be recruited: scientists, researchers, technicians, engineers, even psychologists—because psychology will continue to be a socially necessary instrument, either of servitude or of liberation.

This class, this intelligentsia has been called the new working class. I believe this term is at best premature. They are—and this we should not forget—today the pet beneficiaries of the established system. But they are also at the very source of the glaring contradictions between the liberating capacity of science and its repressive and enslaving use. To activate the repressed and manipulated contradiction, to make it operate as a catalyst of change, that is one of the main tasks of the opposition today. It remains and must remain a political task.

Education is our job, but education in a new sense. Being theory as well as practice, political practice, education today is more than discussion, more than teaching and learning and writing. Unless and until it goes beyond the classroom, until and unless it goes beyond the college, the school, the university, it will remain powerless. Education today must involve the mind *and* the body, reason *and* imagination, the intellectual *and* the instinctual needs, because our entire existence has become the subject/object of politics, of social engineering. I emphasize, it is not a question of making the schools and universities, of making the educational system political. The educational system is political already. I need only remind you of the incredible degree to which (I am speaking of the U.S.) universities are involved in huge research grants (the nature of which you know in many cases) by the government and the various quasi-governmental agencies.

The educational system *is* political, so it is not we who want to politicize the educational system. What we want is a counter-policy against the established policy. And in this sense we must meet this society on its own ground of total mobilization. We must confront indoctrination in servitude with indoctrination in freedom. We must each of us generate in ourselves, and try to generate in others, the instinctual need for a life without fear, without brutality, and without stupidity. And we must see that we can generate the instinctual and intellectual revulsion against the values of an affluence which spreads aggressiveness and suppression throughout the world.

Before I conclude I would like to say my bit about the Hippies. It seems to me a serious phenomenon. If we are talking of the emergence of an instinctual revulsion against the values of the affluent society, I think here is a place where we should look for it. It seems to me that the Hippies, like any nonconformist movement on the left, are split. That there are two parts, or parties, or tendencies. Much of it is mere masquerade and clownery on the private level, and therefore indeed, as Gerassi suggested, completely harmless, very nice and charming in many cases, but that is all there is to it. But that is not the whole story. There is in the Hippies, and especially in such tendencies in the Hippies as the Diggers and the Provos, an inherent political element—perhaps even more so in the U.S. than here. It is the appearance indeed of new instinctual needs and values. This experience is there. There is a new sensibility against efficient and insane reasonableness. There is the refusal to play the rules of a rigid game, a game which one knows is rigid from the beginning, and the revolt against the compulsive cleanliness of puritan morality and the aggression bred by this puritan morality as we see it today in Vietnam among other things.

At least this part of the Hippies, in which sexual, moral and political rebellion are somehow united, is indeed a nonaggressive form of life: a demonstration of an aggressive nonaggressiveness which achieves, at least potentially, the demonstration of qualitatively different values, a transvaluation of values.

All education today is therapy: therapy in the sense of liberating man by all available means from a society in which sooner or later, he is going to be transformed into a brute, even if he doesn't notice it any more. Education in this sense is therapy, and all therapy today is political theory and practice. What kind of political practice? That depends entirely on the situation. It is hardly imaginable that we should discuss this here in detail. I will only remind you of the various possibilities of demonstrations, of finding out flexible modes of demonstration which can cope with the use of institutionalized violence, of boycott, many other things—anything goes which is such that it indeed has a reasonable chance of strengthening the forces of the opposition.

We can prepare for it as educators, as students. Again I say, our role is limited. We are no mass movement. I do not believe that in the near future we will see a mass movement.

I want to add one word about the so-called Third World. I have not spoken of the Third World because my topic was strictly liberation from the affluent society. I agree entirely with Paul Sweezy, that without putting the affluent society in the framework of the Third World it is not understandable. I also believe that here and now our emphasis must be on the advanced industrial societies—not forgetting to do whatever we can and in whatever way we can to support, theoretically and practically, the struggle for liberation in the neocolonial countries which, if again they are not the final force of liberation, at least contribute their share—and it is a considerable share—to the potential weakening and disintegration of the imperialist world system.

Our role as intellectuals is a limited role. On no account should we succumb to any illusions. But even worse than this is to succumb to the wide-spread defeatism which we witness. The preparatory role today is an indispensable role. I believe I am not being too optimistic—I have not in general the reputation of being too optimistic—when I say that we can already see the signs, not only that *They* are getting frightened and worried but that there are far more concrete, far more tangible manifestations of the essential weakness of the system. Therefore, let us continue with whatever we can—no illusions, but even more, no defeatism.

MARX W. WARTOFSKY

Is Science Rational?

Marx W. Wartofsky is Chairman of the Department of Philosophy at Boston University. He is co-editor of *Boston Studies in the Philosophy of Science* and the author of *The Conceptual Foundations of Scientific Thought* (1968). Here he discusses the paradoxical nature of scientific rationality in the context of moral and social commitment.

We face a paradox: scientific rationality, which has liberated man from ignorance, from the whims and oppression of a blind nature, and which has subordinated the earth to man, has become the potential instrument of the self-destruction of the human species. War, pollution and economic oppression are seen as the inevitable spin-off of scientific advance by large sections of the public. The spectre of atomic annihilation and the barbarism of the Hiroshima and Nagasaki bombs are seen as the products of an unrestrained scientific rationality. The recent wave of anti-scientific, anti-rational moods, especially among the young people, threatens a wholesale rejection not simply of the technological fruits of science, but of scientific rationality as well, in favor of one or another version of mysticism, irrationalism, and primitivism, of blood and soil philosophy. Some-how, the argument goes, if we listen to the blood, get back to our roots, and cast out the evil demons of a blind and inhuman rationality, we will save our-selves. On this account of how things stand, the only reasonable thing to do is to reject reason, at least in its scientific forms. And thus the paradox: a reasonable rejection of reason.

We may formulate the question in less paradoxical ways, however. First, it may be asked whether "scientific rationality" is rational. But then we are asking for a definition or a characterization of rationality which is independent of its "scientific" form, against which to test the adequacy of a particular tradition of rationality. Otherwise the question is circular; for if what we mean by rationality is no more than what we mean by science, or by the methodology of science, then the question degenerates into "Is rationality rational?". On the other hand, we may ask the question in still another way: "Is rationality itself adequate to the tasks of human survival? Or is rationality itself a danger to the species? Can

Reprinted by permission of the author; previously unpublished.

it be that the rejection of science may be justified on the grounds that this very rationality itself is no longer a viable instrument of human survival?

Irrationalists, emotivists, voluntarists, fideists, existentialists, among others, have theorized long and hard on the inadequacies of reason, on its limits, and even on its dangers. But I do not intend to enter the lists against such theoretical forms of anti-rationalism here. Rather, what is of more immediate concern is a more popular and less theoretical mood which is specifically anti-scientific, and anti-technological as well.

I would like to cut beneath the appearances and the expressions of this mood, since I see it as a symptom of a deeper malaise. Let me say first off, to make my own stance clear, that I reject *in toto* the rejection of scientific rationality. Yet, I want to acknowledge that the symptoms of malaise may point to a flaw in this rationality itself, to an inadequacy in its conception, and in particular to a disease of its *practice*, which derives from the specific uses to which this scientific rationality can be, and has been put. In short, I want to examine the thesis that science, as liberating reason, can become transformed into its opposite, as repressive reason, and that this transformation threatens the future both of science and of mankind. I also hope to propose a viable alternative.

The thesis I will uphold in this paper is that science *is* reason, that scientific rationality is the most highly adapted and most advanced form of cognition which our species has evolved; and yet, that it stands in danger, for the first time, of becoming dysfunctional or maladaptive. The question I will raise is: How can a functionally adaptive trait, which has been a liberating force throughout its long history and development, become instead dysfunctional, and repressive? The answer to this question, I will suggest, requires scientific study and the development of a theory—namely a theory of rationality—and therefore I pose the question as a research project. Any attempt to answer it in simply rhetorical terms can only result in more of the same sort of pieties or condemnations which characterize the present low level of discussion on this question, and which have been characteristic, in the past, of rationalist or irrationalist or anti-rationalist credos, manifestos and dogmas. In short, I propose a rational framework for the study of science; and thereby, let you know what my prejudices are at the start: That the study of rationality, and the basis for a commitment to reason can only be undertaken by a rational method, and with a commitment to reason. I will argue, finally, that reason, as science, can be used rationally, but also irrationally.

My considerations will be broad, but my intentions are quite specific. We are at the point, in our national development, and in our international relations, where science *policy* becomes the crucial focus of the question of scientific rationality. We are concerned therefore with the rationality of decisions which affect the very practice of science, and which control (or fail to control) the uses of science. Still more crucially, part of this science policy helps to determine the education of new scientists, and the education of the public concerning science. In a sense, we have developed the need for a second order science, one which concerns the practice and uses of science, and we may ask the question whether *this* second-order science is practiced rationally or whether it is, in effect, not scientific at all, or only primitively so at present. Furthermore, we may ask what the credentials should be, for such a science-policy scientists: What are the parameters of the subject? What training is required? What are the criteria of objectivity and rationality in such a science? What are the constraints? Here are the concrete questions which bear on the liberating or repressive uses of reason in our time, and my general considerations are for the sake of these questions.

I will deal first with three questions: (1) What is liberating reason? (2) What is repressive reason? (3) Why choose the framework of evolution and adaptation for the treatment of reason, or as the framework of a theory of rationality?

I take liberating reason as that practice of cognition, of coming to know, which frees man from the bondage to sheer circumstance, and to a dumb environment, by the achievement of mastery over that circumstance and that environment. By mastery, I mean a determination of the will, or an intention, *and* the means of carrying out that determination or intention in practice. I take this to be a condition of freedom, or of liberation, insofar as only such mastery gets us beyond either will-lessness (i.e., being a passive victim, of natural forces), or bondage (i.e., being the passive or helpless object of another's will).[1]

Liberating reason, therefore, is what gives man autonomy, dominion over nature, and the means of effecting his will. One crucial flaw in the earlier theories is that autonomy, dominion, free will (i.e., a will which is in accordance with the "nature" of the willing agent, or which is self-legislating), are all seen as ahistorical features of an abstract human nature. In that sense, these are "ideal type" theories, whose objects are ideal theoretical entities—in this case, "man," "reason," "human nature"—much like the frictionless surfaces and mass points of an idealized mathematical physics. Just as the physicist can work out the consequences of a theory, deductively, given such ideal entities mathematically describable, so too one can work out the deductive consequences of a theory whose ideal entities are logically describable (as "man," or "reason," or "human nature" are, in the theories of rationality, from Plato and Aristotle to Kant and Hegel). The crucial flaw here is not that the types are ideal, but that the account remains merely formal, not yet amenable to that other feature of science, the test of practice, or more accurately, that deliberate test which we call experiment, in which deductive consequences of a theory are checked against practice, or against reality. What ought to count as liberating reason, in historical contexts, is what contributes to man's mastery in a concrete circumstance, in a specific historical period, in a specifically describable way. Only in this way does the formal notion of mastery become a historically relevant notion capable of test in specific contexts, and in concrete instances.

For example, once liberating reason has achieved mastery over some natural or social evil, say, like malaria, it is no longer to be counted as a scientific rational achievement to introduce the cure to a scientifically backward people. Rather, this becomes a piece of technology, or engineering. Reason has done its liberating work, at least in the domain of the theory of the disease. However, it may still have the task of solving a different scientific problem—namely that of the mastery of society, so that the technology of prevention and cure may be optimally pursued, or applied. But this makes the social function and role of science itself a question of scientific theory. It raises questions about the relation of scientific discovery to the technological and social applications of this discovery, or to the social and historical contexts of the discovery itself. In short, it raises questions concerning the history and sociology of science, as integral parts of a theory of the application or uses of scientific theory. And once again, if the study of such questions is not to remain merely formal, it too requires its test in practice—in the social practice of science, guided by a theory (or by theories) of the social role and function of science, and of the genesis of scientific discovery.

This suggests a piggy-backing operation, in which the solution of one scientific question gives rise to another, at a different level. In the example, the theoretical solution of the problem of the causes of malaria gives rise to a second level problem concerning the *uses* or applications of this solution, not simply in the technological terms of social or medical applications of the knowledge, but also in theoretical-scientific terms, that is, in terms of a *theory* of the social application of scientific knowledge. In this way, reason liberates man from ignorance, not simply by intellectual enlightenment—in understanding the causes or nature of things—but from the ignorance of bondage to brute nature, to disease, etc. But there is, in this very process, another aspect of liberation: liberation from bondage to old problems. In the economy of human reason, the solution of a problem frees us to take up new problems, to ramify the domain of rational mastery, or in effect, to enlarge the universe of human dominion. Thus, the other aspect of liberating reason, is that it liberates us, by solving one problem, to discover and take up another; in short, to make scientific progress by the discovery of problems, and of new problems in the bosom of the old.[2]

I have suggested, therefore, that liberating reason is liberating in two ways: Liberation from the blind bondage to natural circumstance, and liberation from bondage to another's will or from one's own will-lessness on the one hand; and on the other, liberation from old problems, by means of their solution, and the generation of new problems. In both of these forms of liberating reason, what is liberated is human activity; or better, what is *achieved* is human activity, by contrast to inhuman or subhuman passivity. That man is free, who is active in the sense of self-active, capable of knowing his will, and of effecting it.

What then is repressive reason? Given the above formulation of liberating reason, one can see, I think, that the seeds of repressive reason lie already in the notion of liberating reason itself. If I am free to effect my will, and if the object of my will is another human being, then my freedom is his unfreedom. I can be active, only at the expense of his passivity. Mastery over nature (as dumb nature, i.e. as natural forces without will) does not turn a "free" nature into an "unfree" nature. Nature is neither free nor unfree, it just *is*. My mastery cannot violate the laws of nature, or suppress them; it can, at best, turn nature's will-less, and intentionless laws to my own, human account, to my will-ful and intentional ends. But I *can* suppress another human being (and by extension, if I were to admit any degree of self-will or purpose to any higher animal, I could be said to suppress that animal) to my will. The domestication of animals, and before it, the hunting of animals, was such an achievement of mastery over other living things to shape them to human needs and human ends; and the piety expressed in hunting rituals, and in animal totemism, among savage peoples is an early form of recognition of the power of deliberate, rational action: The hunted animal, or the animal totem has to be placated for the domination or mastery exerted over it. It is seen as human-like, and the violation of *its* will therefore is empathetically reconstructed in an act of contrition, of guilt felt for violating this will (or spirit). Piety always has this element of atonement and guilt. But let me stretch this point a bit. This guilt (or piety) felt, in this projected animistic form, is in fact the feature of the repressive use of reason, for purposes of mastery. When the object of mastery is another human being, when what is being dominated is the will of another, then reason is split in two: my reason, which liberates me, and gives me mastery, does so by repressing yours, by enslaving you. You have become an object to be manipulated to my ends, no longer a human being, insofar as you are merely a subject to my will. Thus, the seeds of repressive reason are in the repressive uses of my reason, my mastery and domination, over another human being. But if, in fact, my ends, or my will cannot be effected or realized without the means of another person subjected to my will, to carry out these ends, I cannot be free without subjecting you, or another like you— and therefore, like myself.

Now let me take this out of the realm of abstract will and reason, for in this context, the form of this relationship I have been describing has been thoroughly investigated and developed in Hegel's *Phenomenology of Mind,* in his analysis of what he calls *Herrschaft und Knechtschaft*—Domination and Subjection (or in the standard translation, the Master-Slave relation). For what we have here is not yet what I want to call repressive reason, but only its condition—the use of reason in a repressive way, or *repressing* reason. It is still reason, repressing or not, if it is the elaboration, by cognitive means, of a method of effecting one's will or one's intention. But only in the realm of the mastery of man over other men does the question of repressive uses of reason arise. I cannot be said to "repress" nature's will. That's nonsense (nowadays, at any rate; in savage society, however, I wouldn't hold it to have been nonsense, but rather animism—an early prescientific cognitive adaptation to the task of mastering nature, and at that stage, a largely effective one, by contrast to sheer animal behavior). Insofar as rationality is concerned to achieve cognitive mastery over nature, by way of understanding it, discovering its laws, in order to turn them to human account, such rationality feels fully free, or feels itself free to the extent that it doesn't anthropomorphize nature. (In that sense, deanthropomorphizing nature is an exhilarating liberation of reason, in the history of early science.) But in asserting my will over another's, repression enters, and the free rationality of the natural sciences becomes contaminated with human exploitation. In concrete historical circumstances, the very advances of natural science are bound up with social developments, and social uses, in that intricate relation between theory and its social matrix.[3]

But here, reason cannot be considered apart from its moral aspects. It is no longer that "free activity of the mind" in some abstract universe of thought, because its very freedom, in practice, is bound up with another's unfreedom. So the very mastery, which I have characterized as the liberating aspect of reason, is bound up with repression.

Now because my reason is not simply my own, because rationality is early perceived as a species-characteristic (already by the Pre-Socratic Greek philosophers, explicitly, for the *Logos* is what is "common"), the object of my enslavement, of my dominion, has to be defined as a nonrational being, i.e. as nonhuman, so that my mastery can be pursued with impunity, and with a clear mind and heart. For otherwise, the slave's claim to humanity is a claim against my dominion, against my use of him to my own ends. This very division between the human and the subhuman, or nonhuman, dominates our history as an *accepted* theory until the mid-nineteenth century, and we tend to forget, in our present "enlightenment" how pervasive and accepted a view this has been throughout human history. We note with dismay what vestiges of it remain to this day, in racism, in national chauvinism, in male supremacy. But in effect, the social or species-concept of man as by nature a rational animal has been a torn and ruptured concept, because the history of the exploitation of man by man has required that some men be regarded as less than rational, and therefore, as less than human.[4]

This *repressing* reason is the root of *repressive* reason. By repressing reason, I mean reason turned against itself: reason flawed by its self-repression, not simply as a syndrome of individuals, but as a social syndrome; reason kept from its function as a means of mastering nature and of mastering society for human ends, by its use in the domination of one man by another, or of one nation by another, or of one class by another. Here, as in savage society, the guilt which arises from the uses of reason against other human beings results in pieties, with respect to the victims: "We're helping to civilize them," "We're saving their souls," "We're taking care of them, because they are incapable of taking care of themselves"—or better yet, "*We're* liberating *them*." Repressive reason is the use of reason for such purposes: flawed by guilt, embroidered with pieties, but successful thereby, as an instrument for the most effective exploitation of others for my own ends.

But now I am moralizing blatantly, so I'd better stop to give reasons why I should do so. And this brings me to my third question: Why choose the framework of evolution and adaptation for a theory of rationality?

I take reason to be no more mysterious (and perhaps no less mysterious) than any other trait evolved in the course of animal or human evolution. Whatever systematic difficulties there may be with the concept or the theory of adaptation, I propose to accept its broad outlines: Reason, like vertical position, the apposite thumb, and human speech, develops as a trait of human beings, because it functions optimally in human survival. Like these other traits, it is a natural, and not a supernatural or nonnatural phenomenon, and like these other traits, it is a species phenomenon, not an idiosyncratic characteristic. It is as much bound up with social and cultural development as they with it: The two are inseparable, though distinguishable for purposes of analysis.

The natural history of reason is a social history, not simply a biological one, though the biologically necessary conditions have to do with all the usual things one speaks of in these contexts: The development of organs capable of producing speech, the growth of the cerebral cortex, the differentiation of the sensory system, etc. In social terms, one can mention, unsystematically for the occasion, the use of tools, the development of production, of the family, of division of labor—all the happy yet problematic undergrowth of physical and cultural anthropology. Now we haven't even begun the story, but only set the scene. The story, as I see it, is the history of science; but not in the standard senses of some textbooks: i.e., it is neither the history of inventions, nor the stories of great scientists, nor the chronology of discoveries. Rather, it is the *hidden* history of science, which includes all of these, but much more. And this "much more" I would have to characterize as the history of rationality, the prime characteristics of which I take to be the invention of theories, to satisfy curiosity and to explain and to understand what makes us wonder; and thereby to permit us to effectively exercise our will, to predict, plan, deliberate and carry out. Where wishes originate is a deep and complex question. Where needs originate is also difficult, but more apparent. Between needs, and the desires that express them, and the fantasies and actions that fulfill them, there is a complex relation; but in this web, somewhere, is caught the thread of reason. Or perhaps the design of the web *is* reason (if such a thin web can hold such a heavy metaphor). It is our instrumentality for successful action, it is our program of cognitive acquisition, refined in the trial and error of experiment and practice, and shaped into habits, maxims, rules of action, and in its articulate, and criticizable form, put forth as theory. That it is a remarkably adept instrument can be measured by the fact of our species survival and our dominion of the earth.

I choose this evolutionary, and ecological framework, because in it, we can pose the crucial questions I wish to pose: Now that we can reflect, on a sufficiently global scale to be able to conceive of the self-destruction of the species; now that reason has become sufficiently universal to grasp our own limits, in pedestrian, nontheological ways which any child can understand (e.g. that we can blow ourselves up, poison ourselves chemically, starve ourselves off, crowd ourselves to death, on a global scale, in part as the result of deliberate applications of rational science to providing the means of species-suicide, partly by the inadequacy of rational science to foresee the consequences of biological, ecological and demographic proliferation earlier, but mainly as a result of the anarchy of reason—by which I mean its incapacity to become a socialized science), we may now ask the question whether rationality, like some other growth-mechanism, has become dysfunctional and destructive to the future survival of the species. By very rough analogy: has the growth mechanism of rationality become cancerous?

My moral maxims are very simple: Life is good, death is bad. A good life is better than a bad one, and a long life is better than a short one. A meaningful life is best of all. I will not elaborate this litany here. But in this all but specious summary, I see the test of rationality, and with it, of its explicit institutionalized and highest form—science: Does rationality serve to fulfill and enhance these ends for the mass of mankind? Does it pay off, on this sketch of survival goals?

My conclusions can only be programmatic here: Anarchic reason threatens these goals of human survival. That is, reason used simply as the instrumentality of conflicting wills becomes a threat. It is bound to its repressive uses, in this context, and becomes more and more a means of power and exploitation. The limited rationality of science as no more than an instrument of dominion and exploitation by one class or nation over another flaws rationality itself, therefore, and transforms even its liberating features into repressive ones. The alternative, however, is not the abandonment of reason, nor a surrender to mysticism or irrationalism. Such moves are capitulations, for they do not constrain the anarchy of reason, they only enhance it, and effectively yield to its sway. The alternative is a socialized reason—more concretely, a socialized science, or one which takes its *rational* imperative to be its own responsibility for human welfare. I would argue that this is an imperative of reason itself, not simply an "ethical" imperative, as if ethical judgments were somehow *not* linked to reason, or to uses of scientific rationality in determining the uses of science. A bland, or pietistic, or abstract morality, not bound to the rational control of human scientific and technological powers, is a helpless and hopeless morality. But a hard-headed and responsible approach to such moral imperatives as now face us is no more, and no different than a rational use of reason itself. That is, the *uses* of reason can be rational or irrational: social or anarchic. The theory of the social uses of reason is a social-scientific theory. The technology which such a theory deals with is science-policy. But science-policy seen simply as an instrument of national or class domination, or simply as the bureaucratized technique of science-management is itself no more than repressive reason gone professional. Like all technologies, it can become brute uncritical application, divorced from its rational and human contexts. Historically, one may say that even the repressive uses of reason served a function, in the revolutionization of production, in the organization of technology, of society, of nation-states, and all this at a high cost to the exploited, to the victims of war, domination, economic development. But scale has become a major factor here, as elsewhere in nature. On the present scale, of world dominion by science and technology, of world-integration in the political-economy sense, repressive reason is no longer a feasible alternative, since it is this very use of reason which threatens the survival of the species. In this context, liberating reason is the only rational use of reason. Survival itself commands rational ends: a society in which scientific rationality comes under self-control, in which it can be rationally used only to serve the peoples of the earth, not to exploit or suppress them. Any other choice is the suicide of reason by the very instrumentalities which reason itself has developed.

Bibliography

1. This theory of mastery and of self-mastery is fully developed in three classical master-pieces of the science of reason, all of which deal with the relation of reason to will and action: Spinoza's *Ethics,* Immanuel Kant's *Foundations of the Metaphysics of Morals,* and Hegel's *Phenomenology of Mind.* They represent the profoundest and most fully worked out defense of reason in human affairs, and are especially relevant here since, in each of these, the model of human control and self-control is that of science, in its

paradigmatic seventeenth and eighteenth century forms, and also fall within the limitations of that model.

2. There is a formal theory of this particular sequence in cognitive progress, of the interrelationship of problem solving to problem posing, but it is still in its infancy, or in its speculative-mystical stage, encumbered by vagueness on the one hand, and by dogmatic-formalistic attempts at clarity on the other. This formal theory is called dialectic. Its chief innovator, in practice, was Socrates, and its principle formulator was Hegel, who discovered it, in its historical form, in the history of philosophical ideas and theories. In the hands of Marx and Engels, it became an applied theory of social, economic and political development. Contemporary theories of this dialectic, especially in its relation to the growth of scientific knowledge, reappear in more or less degenerate or simplistic forms, e.g. in the works of Popper, Kuhn, and Lakatos. These have the virtue of rediscovery: the theories of cognitive or scientific growth here are not bound to the explicit Hegelian formalisms, and are thus more sensitive to concrete historical and practical contexts. In the recent work of Toulmin, the model of growth is that of evolutionary adaptation of conceptual frameworks. All these variants have a common emphasis on growth or change in science, and exhibit weaker or stronger versions of dialectic.

3. Aristotle, in talking about the development of mathematics in Egypt, already notes that leisure is a condition of scientific theory; that, in effect, one man's leisure means another man's slavery. He writes: "When all such inventions [which aim at utility—M.W.] were already established, the sciences which do not aim at giving pleasure or at the necessities of life were discovered, and first, in the places where men first began to have leisure. This is why the mathematical arts were founded in Egypt; for there the priestly caste was allowed to be at leisure." (*Metaphysics,* 981 b, 19–24).

4. See my "Plato's Republic Revisited: The Dilemma of Philosophy and Politics," in *Human Dignity—This Century and the Next,* ed. R. Gotesky and E. Laszlo (New York: Gordon and Breach, Science Publishers, 1970).

PAUL EHRLICH AND
JOHN P. HOLDREN

Population and Panaceas:
A Technological Perspective

Paul Ehrlich is Professor of Biology at Stanford University in California. He is the author of *The Population Bomb*, the best seller about the population explosion. John P. Holdren is a physicist at the Lawrence Livermore Laboratory of the University of California. In this essay Ehrlich and Holdren discuss the enormous problem that population growth is creating in the world today, particularly the problem of food supply. Technological advances, the authors argue, will not,

Persons interested in the documentation of this article should refer to the numerous footnotes in the original publication.

alone, solve the problem of feeding the world's population. Population control, they argue is a necessity.

Today more than one billion human beings are either undernourished or malnourished, and the human population is growing at a rate of 2% per year. The existing and impending crises in human nutrition and living conditions are well-documented but not widely understood. In particular, there is a tendency among the public, nurtured on Sunday-supplement conceptions of technology, to believe that science has the situation well in hand—that farming the sea and the tropics, irrigating the deserts, and generating cheap nuclear power in abundance hold the key to swift and certain solution of the problem. To espouse this belief is to misjudge the present severity of the situation, the disparate time scales on which technological progress and population growth operate, and the vast complexity of the problems beyond mere food production posed by population pressures. Unfortunately, scientists and engineers have themselves often added to the confusion by failing to distinguish between that which is merely theoretically feasible, and that which is economically and logistically practical.

As we will show here, man's present technology is inadequate to the task of maintaining the world's burgeoning billions, even under the most optimistic assumptions. Furthermore, technology is likely to remain inadequate until such time as the population growth rate is drastically reduced. This is not to assert that present efforts to "revolutionize" tropical agriculture, increase yields of fisheries, desalt water for irrigation, exploit new power sources, and implement related projects are not worthwhile. They may be. They could also easily produce the ultimate disaster for mankind if they are not applied with careful attention to their effects on the ecological systems necessary for our survival (Woodwell, 1967; Cole, 1968). And even if such projects are initiated with unprecedented levels of staffing and expenditures, without population control they are doomed to fall far short. No effort to expand the carrying capacity of the Earth can keep pace with unbridled population growth.

To support these contentions, we summarize briefly the present lopsided balance sheet in the population/food accounting. We then examine the logistics, economics, and possible consequences of some technological schemes which have been proposed to help restore the balance, or, more ambitiously, to permit the maintenance of human populations much larger than today's. The most pertinent aspects of the balance are:

1. The world population reached 3.5 billion in mid-1968, with an annual increment of approximately 70 million people (itself increasing) and a doubling time on the order of 35 years (Population Reference Bureau, 1968).
2. Of this number of people, at least one-half billion are undernourished (deficient in calories or, more succinctly, slowly starving), and approximately an additional billion are malnourished (deficient in particular nutrients, mostly protein) (Borgstrom, 1965; Sukhatme, 1966). Estimates of the number actually perishing annually from starvation begin at 4 million and go up (Ehrlich, 1968) and depend in part on official definitions of starvation which

conceal the true magnitude of hunger's contribution to the death rate (Lelyveld, 1968).
3. Merely to maintain present inadequate nutrition levels, the food requirements of Asia, Africa, and Latin America will, conservatively, increase by 26% in the 10-year period measured from 1965 to 1975 (Paddock and Paddock, 1967). World food production must double in the period 1965–2000 to stay even; it must triple if nutrition is to be brought up to minimum requirements.

Food Production

That there is insufficient additional, good quality agricultural land available in the world to meet these needs is so well documented (Borgstrom, 1965) that we will not belabor the point here. What hope there is must rest with increasing yields on land presently cultivated, bringing marginal land into production, more efficiently exploiting the sea, and bringing less conventional methods of food production to fruition. In all these areas, science and technology play a dominant role. While space does not permit even a cursory look at all the proposals on these topics which have been advanced in recent years, a few representative examples illurstrate our points.

Conventional Agriculture. Probably the most widely recommended means of increasing agricultural yields is through the more intensive use of fertilizers. Their production is straightforward, and a good deal is known about their effective application, although, as with many technologies we consider here, the environmental consequences of heavy fertilizer use are ill understood and potentially dangerous. But even ignoring such problems, we find staggering difficulties barring the implementation of fertilizer technology on the scale required. In this regard the accomplishments of countries such as Japan and the Netherlands are often cited as offering hope to the underdeveloped world. Some perspective on this point is afforded by noting that if India were to apply fertilizer at the per capita level employed by the Netherlands, her fertilizer needs would be nearly half the present world output (United Nations, 1968).

On a more realistic plane, we note that although the goal for nitrogen fertilizer production in 1971 under India's fourth 5-year plan is 2.4 million metric tons (Anonymous, 1968a), Raymond Ewell (who has served as fertilizer production adviser to the Indian government for the past 12 years) suggests that less than 1.1 million metric tons is a more probable figure for that date. Ewell cites poor plant maintenance, raw materials shortages, and power and transportation breakdowns as contributing to continued low production by existing Indian plants. Moreover, even when fertilizer is available, increases in productivity do not necessarily follow. In parts of the underdeveloped world lack of farm credit is limiting fertilizer distribution; elsewhere, internal transportation systems are inadequate to the task. Nor can the problem of educating farmers on the advantages and techniques of fertilizer use be ignored. A recent study (Parikh et al., 1968) of the Intensive Agriculture District Program in the Surat district of Gujarat, India (in which scientific fertilizer use was to have been a major ingredient) notes that "on the whole, the performance of adjoining districts which have similar climate but did not enjoy relative preference of input supply was as good as, if not better than, the programme district. . . . A particularly disheartening feature is that the farm production plans, as yet, do not carry any educative value and have largely failed to convince farmers to use improved practices in their proper combinations."

As a second example of a panacea in the realm of conventional agriculture, mention must be given to the development of new high-yield or high-protein strains of food crops. That such strains have the potential of making a major contribution to the food supply of the world is beyond doubt, but this potential is limited in contrast to the potential for population growth, and will be realized too slowly to have anything but a small impact on the immediate crisis. There are major difficulties impeding the wide-spread use of new high-yield grain varieties. Typically, the new grains require high fertilizer inputs to realize their full potential, and thus are subject to all the difficulties mentioned above. Some other problems are identified in a recent address by Lester R. Brown, administrator of the International Agricultural Development Service: the limited amount of irrigated land suitable for the new varieties, the fact that a farmer's willingness to innovate fluctuates with the market prices (which may be driven down by high-yield crops), and the possibility of ticups at market facilities inadequate for handling increased yields

Perhaps even more important, the new grain varieties are being rushed into production without adequate field testing, so that we are unsure of how resistant they will be to the attacks of insects and plant diseases. William Paddock has presented a plant pathologist's view of the crash programs to shift to new varieties (Paddock, 1967). He describes India's dramatic program of planting improved Mexican wheat, and continues: "Such a rapid switch to a new variety is clearly understandable in a country that tottered on the brink of famine. Yet with such limited testing, one wonders what unknown pathogens await a climatic change which will give the environmental conditions needed for their growth." Introduction of the new varieties creates enlarged monocultures of plants with essentially unknown levels of resistance to disaster. Clearly, one of the prices that is paid for higher yield is a higher risk of widespread catastrophe. And the risks are far from local: since the new varieties require more "input" of pesticides (with all their deleterious ecological side effects), these crops may ultimately contribute to the defeat of other environment-related panaceas, such as extracting larger amounts of food from the sea.

A final problem must be mentioned in connection with these strains of food crops. In general, the hungriest people in the world are also those with the most conservative food habits. Even rather minor changes, such as that from a rice variety in which the cooked grains stick together to one in which the grains fall apart, may make new foods unacceptable. It seems to be an unhappy fact of human existence that people would rather starve than eat a nutritious substance which they do not recognize as food.

Beyond the economic, ecological, and sociological problems already mentioned in connection with high-yield agriculture, there is the overall problem of time. We need time to breed the desired characteristics of yield and hardiness into a vast array of new strains (a tedious process indeed), time to convince farmers that it is necessary that they change their time-honored ways of cultivation, and time to convince hungry people to change the staples of their diet. The Paddocks give 20 years as the "rule of thumb" for a new technique or plant variety to progress from conception to substantial impact on farming (Paddock and Paddock, 1967). They write: "It is true that a *massive* research attack on the problem could bring some striking results in less than 20 years. But I do not find such an attack remotely contemplated in the thinking of those officials capable of initiating it." Promising as high-yield agriculture may be, the funds, the personnel, the ecological expertise, and the necessary years are unfortunately not at our disposal. Fulfillment of the promise will come too late for many of the world's starving millions, if it comes at all.

Bring More Land Under Cultivation. The most frequently mentioned means of bringing new land into agricultural production are farming the tropics and irrigating arid and semiarid regions. The former, although widely discussed in optimistic terms, has been tried for years with incredibly poor results, and even recent experiments have not been encouraging. One essential difficulty is the unsuitability of tropical soils for supporting typical foodstuffs instead of jungles (McNeil, 1964; Paddock and Paddock, 1964). Also, "the tropics" are a biologically more diverse area than the temperate zones, so that farming technology developed for one area will all too often prove useless in others. We shall see that irrigating the deserts, while more promising, has serious limitations in terms of scale, cost, and lead time.

The feasible approaches to irrigation of arid lands appear to be limited to large-scale water projects involving dams and transport in canals, and desalination of ocean and brackish water. Supplies of usable ground water are already badly depleted in most areas where they are accessible, and natural recharge is low enough in most arid regions that such supplies do not offer a long-term solution in any case. Some recent statistics will give perspective to the discussion of water projects and desalting which follows. In 1966, the United States was using about 300 billion gal of water per day, of which 135 billion gal were consumed by agriculture and 165 billion gal by municipal and industrial users (Sporn, 1966). The bulk of the agricultural water cost the farmer from 5 to 10 cents/1000 gal; the highest price paid for agricultural water was 15 cents/1000 gal. For small industrial and municipal supplies, prices as high as 50 to 70 cents/1000 gal were prevalent in the U.S. arid regions, and some communities in the Southwest were paying on the order of $1.00/1000 gal for "project" water. The extremely high cost of the latter stems largely from transportation costs, which have been estimated at 5 to 15 cents/1000 gal per 100 miles (International Atomic Energy Agency, 1964).

We now examine briefly the implications of such numbers in considering the irrigation of the deserts. The most ambitious water project yet conceived in this country is the North American Water and Power Alliance, which proposes to distribute water from the great rivers of Canada to thirsty locations all over the

United States. Formidable political problems aside (some based on the certainty that in the face of expanding populations, demands for water will eventually arise at the source), this project would involve the expenditure of $100 billion in construction costs over a 20-year completion period. At the end of this time, the yield to the United States would be 69 million acre feet of water annually (Kelly, 1966), or 63 billion gal per day. If past experience with massive water projects is any guide, these figures are over-optimistic, but if we assume they are not, it is instructive to note that this monumental undertaking would provide for an increase of only 21% in the water consumption of the United States, during a period in which the population is expected to increase by between 25 and 43% (U.S. Dept. of Commerce, 1966). To assess the possible contribution to the *world* food situation, we assume that all this water could be devoted to agriculture, although extrapolation of present consumption patterns indicates that only about one-half would be. Then using the rather optimistic figure of 500 gal per day to grow the food to feed one person, we find that this project could feed 126 million additional people. Since this is less than 8% of the projected world population growth during the construction period (say 1970 to 1990), it should be clear that even the most massive water projects can make but a token contribution to the solution of the world food problem in the long term. And in the crucial short term—the years preceding 1980—*no* additional people will be fed by projects still on the drawing board today.

In summary, the cost is staggering, the scale insufficient, and the lead time too long. Nor need we resort to such speculation about the future for proof of the failure of technological "solutions" in the absence of population control. The highly touted and very expensive Aswan Dam project, now nearing completion, will ultimately supply food (at the present miserable diet level) for less than Egypt's population growth during the time of construction (Borgstrom, 1965; Cole, 1968). Furthermore, its effect on the fertility of the Nile Delta may be disastrous, and, as with all water projects of this nature, silting of the reservoir will destroy the gains in the long term (perhaps in 100 years).

Desalting for irrigation suffers somewhat similar limitations. The desalting plants operational in the world today produce water at individual rates of 7.5 million gal/day and less, at a cost of 75 cents/1000 gal and up, the cost increasing as the plant size decreases (Bender, 1969). The most optimistic firm proposal which anyone seems to have made for desalting with present or soon-to-be available technology is a 150 million gal per day nuclear-powered installation studied by the Bechtel Corp. for the Los Angeles Metropolitan Water District. Bechtel's early figures indicated that water from this complex would be available at the site for 27–28 cents/1000 gal (Galstann and Currier, 1967). However, skepticism regarding the economic assumptions leading to these figures (Milliman, 1966) has since proven justified—the project was shelved after spiralling construction cost estimates indicated an actual water cost of 40–50 cents/1000 gal. Use of even the original figures, however, bears out our contention that the *most* optimistic assumptions do not alter the verdict that technology is losing the food/population battle. For 28 cents/1000 gal is still approximately twice the cost which farmers have hitherto been willing or able to pay for irrigation water.

If the Bechtal plant had been intended to supply agricultural needs, which it was not, one would have had to add to an already unacceptable price the very substantial cost of transporting the water inland.

Significantly, studies have shown that the economies of scale in the distillation process are essentially exhausted by a 150 million gal per day plant (International Atomic Energy Agency, 1964). Hence, merely increasing desalting capacity further will not substantially lower the cost of the water. On purely economic grounds, then, it is unlikely that desalting will play a major role in food production by conventional agriculture in the short term. Technological "breakthroughs" will presumably improve this outlook with the passage of time, but world population growth will not wait.

Desalting becomes more promising if the high cost of the water can be offset by increased agricultural yields per gallon and, perhaps, use of a single nuclear installation to provide power for both the desalting and profitable on-site industrial processes. This prospect has been investigated in a thorough and well-documented study headed by E.A. Mason (Oak Ridge National Laboratory, 1968). The result is a set of preliminary figures and recommendations regarding nuclear-powered "agro-industrial complexes" for arid and semi-arid regions, in which desalted water and fertilizer would be produced for use on an adjacent, highly efficient farm. In underdeveloped countries incapable of using the full excess power output of the reactor, this energy would be consumed in on-site production of industrial materials for sale on the world market. Both near-term (10 years hence) and far-term (20 years hence) technologies are considered, as are various mixes of farm and industrial products. The representative near-term case for which a detailed cost breakdown is given involves a seaside facility with a desalting capacity of 1 billion gal/day, a farm size of 320,000 acres, and an industrial electric power consumption of 1585 Mw. The initial investment for this complex is estimated at $1.8 billion, and annual operating costs at $236 million. If both the food and the industrial materials produced were sold (as opposed to giving the food, at least, to those in need who could not pay), the estimated profit for such a complex, before subtracting financing costs, would be 14.6%.

The authors of the study are commendably cautious in outlining the assumptions and uncertainties upon which these figures rest. The key assumption is that 200 gal/day of water will grow the 2500 calories required to feed one person. Water/calorie ratios of this order or less have been achieved by the top 20% of farmers specializing in such crops as wheat, potatoes, and tomatoes; but more water is required for needed protein-rich crops such as peanuts and soybeans. The authors identify the uncertainty that crops usually raised separately can be grown together in tight rotation on the same piece of land. Problems of water storage between periods of peak irrigation demand, optimal patterns of crop rotation, and seasonal acreage variations are also mentioned. These "ifs" and assumptions, and those associated with the other technologies involved, are unfortunately often omitted when the results of such painstaking studies are summarized for more popular consumption (Anonymous, 1968b, 1968c). The result is the perpetuation of the public's tendency to confuse feasible and

available, to see panaceas where scientists in the field concerned see only potential, realizable with massive infusions of time and money.

It is instructive, nevertheless, to examine the impact on the world food problem
which the Oak Ridge complexes might have if construction were to begin today,
and if all the assumptions about technology 10 years hence were valid *now*. At
the industrial-agricultural mix pertinent to the sample case described above, the
food produced would be adequate for just under 3 million people. This means
that 23 such plants per year, at a cost of $41 billion, would have to be put in
operation merely to keep pace with world population growth, to say nothing of
improving the sub-standard diets of between one and two billion members of
the present population. (Fertilizer production beyond that required for the on-site
farm is of course a contribution in the latter regard, but the substantial additional
costs of transporting it to where it is needed must then be accounted for.)
Since approximately 5 years from the start of construction would be required to
put such a complex into operation, we should commence work on at least 125
units post-haste, and begin at least 25 per year thereafter. If the technology *were*
available now, the investment in construction over the next 5 years, prior to
operation of the first plants, would be $315 billion—about 20 times the total U.S.
foreign aid expenditure during the past 5 years. By the time the technology *is*
available the bill will be much higher, if famine has not "solved" the problem
for us.

This example again illustrates that scale, time, and cost are all working against
technology in the short term. And if population growth is not decelerated, the
increasing severity of population-related crises will surely neutralize the technological improvements of the middle and long terms.

Other Food Panaceas. "Food from the sea" is the most prevalent "answer" to
the world food shortage in the view of the general public. This is not surprising,
since estimates of the theoretical fisheries productivity of the sea run up to some
50–100 times current yields (Schmitt, 1965; Christy and Scott, 1965). Many
practical and economic difficulties, however, make it clear that such a figure will
never be reached, and that it will not even be approached in the foreseeable
future. In 1966, the annual fisheries harvest was some 57 million metric tons
(United Nations, 1968). A careful analysis (Meseck, 1961) indicates that this
might be increased to a world production of 70 million metric tons by 1980. If
this gain were realized, it would represent (assuming no violent change in population growth patterns) a small per capita *loss* in fisheries yield.

Both the short- and long-term outlooks for taking food from the sea are
clouded by the problems of overexploitation, pollution (which is generally
ignored by those calculating potential yields), and economics. Solving these
problems will require more than technological legerdemain; it will also require
unprecedented changes in human behavior, especially in the area of international
cooperation. The unlikelihood that such cooperation will come about is reflected
in the recent news (Anonymous, 1968d) that Norway has dropped out of the
whaling industry because overfishing has depleted the stock below the level at
which it may economically be harvested. In that industry, international controls

were tried—and failed. The sea is, unfortunately, a "commons" (Hardin, 1968), and the resultant management problems exacerbate the biological and technical problems of greatly increasing our "take." One suspects that the return per dollar poured into the sea will be much less than the corresponding return from the land for many years, and the return from the land has already been found wanting.

Synthetic foods, protein culture with petroleum, saline agriculture, and weather modification all may hold promise for the future, but all are at present expensive and available only on an extremely limited scale. The research to improve this situation will also be expensive, and, of course, time-consuming. In the absence of funding, it will not occur at all, a fact which occasionally eludes the public and the Congress.

Domestic and Industrial Water Supplies

The world has water problems, even exclusive of the situation in agriculture. Although total precipitation should in theory be adequate in quantity for several further doublings of population, serious shortages arising from problems of quality, irregularity, and distribution already plague much of the world. Underdeveloped countries will find the water needs of industrialization staggering: 240,000 gal of water are required to produce a ton of newsprint; 650,000 gal, to produce a ton of steel (International Atomic Energy Agency, 1964). Since maximum acceptable water costs for domestic and industrial use are higher than for agriculture, those who can afford it are or soon will be using desalination (40–100+ cents/1000 gal) and used-water renovation (54–57 cents/1000 gal [Ennis, 1967]). Those who cannot afford it are faced with allocating existing supplies between industry and agriculture, and as we have seen, they must choose the latter. In this circumstance, the standard of living remains pitifully low. Technology's only present answer is massive externally-financed complexes of the sort considered above, and we have already suggested there the improbability that we are prepared to pay the bill rung up by present population growth.

The widespread use of desalted water by those who *can* afford it brings up another problem only rarely mentioned to date, the disposal of the salts. The product of the distillation processes in present use is a hot brine with salt concentration several times that of seawater. Both the temperature and the salinity of this effluent will prove fatal to local marine life if it is simply exhausted to the ocean. The most optimistic statement we have seen on this problem is that "*smaller plants* (our emphasis) at seaside locations may return the concentrated brine to the ocean if proper attention is paid to the design of the outfall, and to the effect on the local marine ecology" (McIlhenny, 1966). The same writer identifies the major economic uncertainties connected with extracting the salts for sale (to do so is straightforward, but often not profitable). Nor can one simply evaporate the brine and leave the residue in a pile—the 150 million gal/day plant mentioned above would produce brine bearing 90 million lb. of salts daily (based on figures by Parker, 1966). This amount of salt would cover over 15 acres to a depth of one foot. Thus, every year a plant of the billion gallon per day, agro-industrial complex size would produce a pile of salt over 52 ft deep and covering a square mile. The high winds typical of coastal deserts would seriously aggravate the associated soil contamination problem.

Energy. Man's problems with energy supply are more subtle than those with food and water: we are not yet running out of energy, but we are being forced to use it faster than is probably healthy. The rapacious depletion of our fossil fuels is already forcing us to consider more expensive mining techniques to gain access to lower-grade deposits, such as the oil shales, and even the status of our high-grade uranium ore reserves is not clearcut.

A widely held misconception in this connection is that nuclear power is "dirt cheap," and as such represents a panacea for developed and underdeveloped nations alike. To the contrary, the largest nuclear-generating stations now in operation are just competitive with or marginally superior to modern coal-fired plants of comparable size (where coal is not scarce); at best, both produce power for on the order of 4–5 mills (tenths of a cent) per kilowatt-hour. Smaller nuclear units remain less economical than their fossil-fueled counterparts. Under-developed countries can rarely use the power of the larger plants. Simply speak-ing, there are not enough industries, appliances, and light bulbs to absorb the output, and the cost of industrialization and modernization exceeds the cost of the power required to sustain it by orders of magnitude, regardless of the source of the power. (For example, one study noted that the capital requirement to consume the output of a 70,000 kilowatt plant—about $1.2 million worth of electricity per year at 40% utilization and 5 mills/kwh—is $111 million per year if the power is consumed by metals industries, $270 million per year for petro-leum product industries [E.A. Mason, 1957].) Hence, at least at present, only those underdeveloped countries which are short of fossil fuels or inexpensive means to transport them are in particular need of nuclear power.

Prospects for major reductions in the cost of nuclear power in the future hinge on the long-awaited breeder reactor and the still further distant thermonuclear reactor. In neither case is the time scale nor the ultimate cost of energy a matter of any certainty. The breeder reactor, which converts more nonfissile uranium (^{238}U) or thorium to fissionable material than it consumes as fuel for itself, effectively extends our nuclear fuel supply by a factor of approximately 400 (Cloud, 1968). It is not expected to become competitive economically with con-ventional reactors until the 1980's (Bump, 1967). Reductions in the unit energy cost beyond this date are not guaranteed, due both to the probable continued high capital cost of breeder reactors and to increasing costs for the ore which the breeders will convert to fuel. In the latter regard, we mention that although crushing granite for its few parts per million of uranium and thorium is possible in theory, the problems and cost of doing so are far from resolved. It is too soon to predict the costs associated with a fusion reactor (few who work in the field will predict whether such a device will work at all within the next 15–20 years). One guess puts the unit energy cost at something over half that for a coal or fission power station of comparable size (Mills, 1967), but this is pure speculation.

Quite possibly the major benefit of controlled fusion will again be to extend the energy supply rather than to cheapen it.

A second misconception about nuclear power is that it can reduce our dependence on fossil fuels to zero as soon as that becomes necessary or desirable. In fact, nuclear power plants contribute only to the electrical portion of the energy budget; and in 1960 in the United States, for example, electrical energy comprised only 19% of the total energy consumed (Sporn, 1963). The degree to which nuclear fuels can postpone the exhaustion of our coal and oil depends on the extent to which that 19% is enlarged. The task is far from a trivial one, and will involve transitions to electric or fuel-cell powered transportation, electric heating, and electrically powered industries. It will be extremely expensive.

Nuclear energy, then, is a panacea neither for us nor for the underdeveloped world. It relieves, but does not remove, the pressure on fossil supplies; it provides reasonably-priced power where these fuels are not abundant; it has substantial (but expensive) potential in intelligent applications such as that suggested in the Oak Ridge study discussed above; and it shares the propensity of fast-growing technology to unpleasant side effects (Novick, 1969). We mention in the last connection that, while nuclear power stations do not produce conventional air pollutants, their radioactive waste problems may in the long run prove a poor trade. Although the AEC seems to have made a good case for solidification and storage in salt mines of the bulk of the radioactive fission products (Blanko et al., 1967), a number of radioactive isotopes are released to the environment, and in some areas such isotopes have already turned up in potentially harmful concentrations (Curtis and Hogan, 1969). Project order of magnitude increases in nuclear power generation will seriously aggravate this situation. Although it has frequently been stated that the eventual advent of fusion reactors will free us from such difficulties, at least one authority, F.L. Parker, takes a more cautious view. He contends that the large inventory of radioactive tritium in early fusion reactors will require new precautions to minimize emissions (Parker, 1968).

A more easily evaluated problem is the tremendous quantity of waste heat generated at nuclear installations (to say nothing of the usable power output, which, as with power from whatever source, must also ultimately be dissipated as heat). Both have potentially disastrous effects on the local and world ecological and climatological balance. There is no simple solution to this problem, for, in general, "cooling" only moves heat; it does not *remove* it from the environment viewed as a whole. Moreover, the Second Law of Thermodynamics puts a ceiling on the efficiency with which we can do even this much, i.e., concentrate and transport heat. In effect, the Second Law condemns us to aggravate the total problem by generating still *more* heat in any machinery we devise for local cooling (consider, for example, refrigerators and air conditioners).

The only heat which actually leaves the whole system, the Earth, is that which can be radiated back into space. This amount steadily is being diminished as combustion of hydrocarbon fuels increases the atmospheric percentage of CO_2 which has strong absorption bands in the infrared spectrum of the outbound heat energy. (Hubbert, 1962, puts the increase in the CO_2 content of the atmosphere at 10% since 1900.) There is, of course, a competing effect in the

Earth's energy balance, which is the increased reflectivity of the upper atmosphere to incoming sunlight due to other forms of air pollution. It has been estimated, ignoring both these effects, that man risks drastic (and perhaps catastrophic) climatological change if the amount of heat he dissipates in the environment on a global scale reaches 1% of the solar energy absorbed and reradiated at the Earth's surface (Rose and Clark, 1961). At the present 5% rate of increase in world energy consumption, this level will be reached in less than a century, and in the immediate future the direct contribution of man's power consumption will create serious local problems. If we may safely rule out circumvention of the Second Law or the divorce of energy requirements from population size, this suggests that, whatever science and technology may accomplish, population growth must be stopped.

Transportation

We would be remiss in our offer of a technological perspective on population problems without some mention of the difficulties associated with transporting large quantities of food, material, or people across the face of the Earth. While our grain exports have not begun to satisfy the hunger of the underdeveloped world, they already have taxed our ability to transport food in bulk over large distances. The total amount of goods of *all* kinds loaded at U.S. ports for external trade was 158 million metric tons in 1965 (United Nations, 1968). This is coincidentally the approximate amount of grain which would have been required to make up the dietary shortages of the underdeveloped world in the same year (Sukhatme, 1966). Thus, if the United States *had* such an amount of grain to ship, it could be handled only by displacing the entirety of our export trade. In a similar vein, the gross weight of the fertilizer, in excess of present consumption, required in the underdeveloped world to feed the additional population there in 1980 will amount to approximately the same figure—150 million metric tons (Sukhatme, 1966). Assuming that a substantial fraction of this fertilizer, should it be available at all, will have to be shipped about, we had best start building freighters! These problems, and the even more discouraging one of internal transportation in the hungry countries, coupled with the complexities of international finance and marketing which have hobbled even present aid programs, complete a dismal picture of the prospects for "external" solutions to ballooning food requirements in much of the world.

Those who envision migration as a solution to problems of food, land, and water distribution not only ignore the fact that the world has no promising place to put more people, they simply have not looked at the numbers of the transportation game. Neglecting the fact that migration and relocation costs would probably amount to a minimum of several thousand dollars per person, we find, for example, that the entire long-range jet transport fleet of the United States (about 600 planes [Molloy, 1968] with an average capacity of 150), averaging two round trips per week, could transport only about 9 million people per year from India to the United States. This amounts to about 75% of that country's annual population *growth* (Population Reference Bureau, 1968).

Ocean liners and transports, while larger, are less numerous and much slower, and over long distances could not do as well. Does anyone believe, then, that we are going to compensate for the world's population growth by sending the excess to the planets? If there were a place to go on Earth, financially and logistically we could not send our surplus there.

Conclusion

We have not attempted to be comprehensive in our treatment of population pressures and the prospects of coping with them technologically; rather, we hope simply to have given enough illustrations to make plausible our contention that technology, without population control, cannot meet the challenge. It may be argued that we have shown only that any one technological scheme taken individually is insufficient to the task at hand, whereas *all* such schemes applied in parallel might well be enough. We would reply that neither the commitment nor the resources to implement them all exists, and indeed that many may prove mutually exclusive (e.g. harvesting algae may diminish fish production).

Certainly, an optimum combination of efforts exists in theory, but we assert that no organized attempt to find it is being made, and that our examination of its probable eventual constituents permits little hope that even the optimum will suffice. Indeed, after a far more thorough survey of the prospects than we have attempted here, the President's Science Advisory Committee Panel on the world food supply concluded (PSAC, 1967): "The solution of the problem that will exist after about 1985 *demands* that programs of population control be initiated now." We most emphatically agree, noting that "now" was 2 years ago!

Of the problems arising out of population growth in the short, middle, and long terms, we have emphasized the first group. For mankind must pass the first hurdles—food and water for the next 20 years—to be granted the privilege of confronting such dilemmas as the exhaustion of mineral resources and physical space later. Furthermore, we have not conveyed the extent of our concern for the environmental deterioration which has accompanied the population explosion, and for the catastrophic ecological consequences which would attend many of the proposed technological "solutions" to the population/food crisis. Nor have we treated the point that "development" of the rest of the world to the standards of the West probably would be lethal ecologically (Ehrlich and Ehrlich, 1970). For even if such grim prospects are ignored, it is abundantly clear that in terms of cost, lead time, and implementation on the scale required, technology without population control will be too little and too late.

What hope there is lies not, of course, in abandoning attempts at technological solutions; on the contrary, they must be pursued at unprecedented levels, with unprecedented judgment, and above all with unprecedented attention to their ecological consequences. We need dramatic programs now to find ways of ameliorating the food crisis—to buy time for humanity until the inevitable delay accompanying population control efforts has passed. But it cannot be emphasized enough that if the population control measures are *not* initiated immediately and effectively, all the technology man can bring to bear will not fend off the misery to come. Therefore, confronted as we are with limited resources of time and money, we must consider carefully what fraction of our effort should be applied to the cure of the disease itself instead of to the temporary relief of the symptoms. We should ask, for example, how many vasectomies could be performed by a program funded with the 1.8 billion dollars required to build a single nuclear agro-industrial complex, and what the relative impact on the problem would be in both the short and long terms.

The decision for population control will be opposed by growth-minded economists and businessmen, by nationalistic statesmen, by zealous religious leaders, and by the myopic and well-fed of every description. It is therefore incumbent on all who sense the limitations of technology and the fragility of the environmental balance to make themselves heard above the hollow, optimistic chorus—to convince society and its leaders that there is no alternative but the cessation of our irresponsible, all-demanding, and all-consuming population growth.

Recommended Readings

Bottomore, T.B., *Elites and Society* (New York: Basic Books, 1964).

Edel, A., "Science and the Structure of Ethics," vol. III of *International Encyclopedia of Unified Science* (Chicago: Univ. of Chicago Press, 1961).

Ferkiss, V., *Technological Man* (New York: George Braziller, 1969).

Galbraith, J.K., *The New Industrial State* (Boston: Houghton Mifflin, 1967).

Greenberg, D.S., *The Politics of Pure Science* (New York: The New American Library, 1967).

Haldane, J.B.S., *The Marxist Philosophy and the Sciences* (New York: Random House, 1939).

Hall, E.W., *Modern Science and Human Values* (Princeton, New Jersey: Van Nostrand, 1956).

Irving, J.A., *Science and Value* (Toronto: Ryerson Press, 1952).

Nieburg, H.L., *In the Name of Science* (Chicago: Quadrangle Books, 1966).

Polanyi, M., *The Logic of Liberty* (Chicago: Univ. of Chicago Press, 1951).

Science, Technology, and the Environment

Science, Technology, and the Environment

4

Introduction

The effects of pollution are predictable. The exquisite balance of natural cycles in which one creature's "waste" is another's livelihood is upset. What happens? First of all, there is a general simplification of the environment. There are fewer species because those more susceptible to pollutants are destroyed, while resistant ones survive, and often thrive. Taller plants, trees, go first, and with them a multitude of habitats for birds, insects, vines, ferns, etc. Nutrients are lost, energy is wasted. The survivors are the organisms which can get by in an environment stripped of its organic richness. As more people make more pollution, we head for an environment of weeds, rats, and roaches.

This is essentially the view of those who have made an unbiased, scientific study of the pollution problem. There is much less agreement about the causes of the problem. Causes can be approached from several directions. We may deal more easily with the *material* causes; too many people consuming and digesting too much food, too much production of too much sewage; too many power plants providing for too many factories, washing machines, dishwashers, air conditioners, and so on. For all this consumption, someone must pay. Often those who really pay, in terms of comfort and life quality, are not those who consume the most. We see from the readings in this chapter that the greatest pollutors—we ourselves—are a minority of the world's population. Let us take an example from our own society. In New York City, government offices, theatres, restaurants and apart-ment houses (for the middle and upper income inhabitants) are air-conditioned. These air conditioners cool the insides of the buildings and vent hot exhaust outside. The penthouses stay serene and quiet, while the ghettoes become hot and agitated. Although we can see and explain the material causes of pollution, the social and spiritual causes are more difficult to define.

This section examines the problem and suggests some possible causes, material and social. The suggested solutions to the problem depend on whether the author sees the root causes as basically socioeconomic, or simply as a new technological challenge. Thus, Huxley and Commoner tell us that a solution can come only through a radical reorientation of social goals. While they do not suggest that nothing can be done about environmental destruction, they are not optimistic about current progress in effecting the kinds of far-reaching changes they feel are needed.

On the other hand, Diamond and Lane, while warning about the dire conse-quences of inaction on the environmental crisis, take an optimistic view about the ability of our society, as it exists, to clean itself up. We don't really have to worry, they seem to tell us, our institutions can do the job. All we need is a little extra push from the already aroused public. There is no need to give up the profit motive all together, just be a little less greedy. We must all get together, like Nader's Raiders, and depollute.

Just how far we need to go in finding our "Environmental Ethic" can be seen from the stark reality of "Land War." It is impossible not to reflect upon the difficulties involved in trying to halt this sort of pollution through existing avenues of public communication and control. It is, after all, the very institution—the government—in which we place our hope and trust for redress that wages war against the land.

Finally, we come to the question of what changes we want to make in our environment, if, in fact, we can succeed in curing the disease of pollution. René Dubos gently and rationally shows us how to approach the question.

ALDOUS HUXLEY

The Politics of Ecology:
The Question of Survival

Aldous Huxley was an internationally-known British novelist and essayist. Among his most widely read books is *Brave New World* (1939). In this article, Huxley describes how control of the death rate in underdeveloped countries has far outstripped control of the birthrate, and indicates what the consequences of the resulting population increase will be, both for the underdeveloped countries and for the Western democracies. He calls for the development of a global politics which disavows chauvinistic nationalism and which has its ideological basis in biological reality.

. . . On the biological level, advancing science and technology have set going a revolutionary process that seems to be destined for the next century at least, perhaps for much longer, to exercise a decisive influence upon the destinies of all human societies and their individual members. In the course of the last fifty years extremely effective methods for lowering the prevailing rates of infant and adult mortality were developed by Western scientists. These methods were very simple and could be applied with the expenditure of very little money by very small numbers of not very highly trained technicians. For these reasons, and because everyone regards life as intrinsically good and death as intrinsically bad, they were in fact applied on a worldwide scale. The results were spectacular. In the past, high birth rates were balanced by high death rates. Thanks to science, death rates have been halved but, except in the most highly industrialized, contraceptive-using countries, birth rates remain as high as ever. An enormous and accelerating increase in human numbers has been the inevitable consequence.

At the beginning of the Christian era, so demographers assure us, our planet supported a human population of about two hundred and fifty millions. When the Pilgrim Fathers stepped ashore, the figure had risen to about five hundred millions. We see, then, that in the relatively recent past it took sixteen hundred years for the human species to double its numbers. Today world population

"The Politics of Ecology: The Question of Survival" by Aldous Huxley, 1964. Reprinted, with permission, from an Occasional Paper, "The Politics of Ecology," a publication of The Center for the Study of Democratic Institutions in Santa Barbara, California.

stands at three thousand millions. By the year 2000, unless something appallingly bad or miraculously good should happen in the interval, six thousand millions of us will be sitting down to breakfast every morning. In a word, twelve times as many people are destined to double their numbers in one-fortieth of the time.

This is not the whole story. In many areas of the world human numbers are increasing at a rate much higher than the average for the whole species. In India, for example, the rate of increase is now 2.3 percent per annum. By 1990 its four hundred and fifty million inhabitants will have become nine hundred million inhabitants. A comparable rate of increase will raise the population of China to the billion mark by 1980. In Ceylon, in Egypt, in many of the countries of South and Central America, human numbers are increasing at an annual rate of 3 percent. The result will be a doubling of their present populations in approximately twenty-three years.

On the social, political, and economic levels, what is likely to happen in an underdeveloped country whose people double themselves in a single generation, or even less? An underdeveloped society is a society without adequate capital resources (for capital is what is left over after primary needs have been satisfied, and in underdeveloped countries most people never satisfy their primary needs); a society without a sufficient force of trained teachers, administrators, and technicians; a society with few or no industries and few or no developed sources of industrial power; a society, finally, with enormous arrears to be made good in food production, education, road building, housing, and sanitation. A quarter of a century from now, when there will be twice as many of them as there are today, what is the likelihood that the members of such a society will be better fed, housed, clothed, and schooled than at present? And what are the chances in such a society for the maintenance, if they already exist, or the creation, if they do not exist, of democratic institutions?

Not long ago Mr. Eugene Black, the former president of the World Bank, expressed the opinion that it would be extremely difficult, perhaps even impossible, for an underdeveloped country with a very rapid rate of population increase to achieve full industrialization. All its resources, he pointed out, would be absorbed year by year in the task of supplying, or not quite supplying, the primary needs of its new members. Merely to stand still, to maintain its current subhumanly inadequate standard of living, will require hard work and the expenditure of all the nation's available capital. Available capital may be increased by loans and gifts from abroad; but in a world where the industrialized nations are involved in power politics and an increasingly expensive armament race, there will never be enough foreign aid to make much difference. And even if the loans and gifts to underdeveloped countries were to be substantially increased, any resulting gains would be largely nullified by the uncontrolled population explosion.

The situation of these nations with such rapidly increasing populations reminds one of Lewis Carroll's parable in *Through the Looking Glass,* where Alice and the Red Queen start running at full speed and run for a long time until Alice is completely out of breath. When they stop, Alice is amazed to see that they are still at their starting point. In the looking glass world, if you wish to retain your present position, you must run as fast as you can. If you wish to get ahead, you must run at least twice as fast as you can.

If Mr. Black is correct (and there are plenty of economists and demographers who share his opinion), the outlook for most of the world's newly independent and economically nonviable nations is gloomy indeed. To those that have shall be given. Within the next ten or twenty years, if war can be avoided, poverty will almost have disappeared from the highly industrialized and contraceptive-using societies of the West. Meanwhile, in the underdeveloped and uncontrolledly breeding societies of Asia, Africa, and Latin America the condition of the masses (twice as numerous, a generation from now, as they are today) will have become no better and may even be decidedly worse than it is at present. Such a decline is foreshadowed by current statistics of the Food and Agriculture Organization of the United Nations. In some underdeveloped regions of the world, we are told, people are somewhat less adequately fed, clothed, and housed than were their parents and grandparents thirty and forty years ago. And what of elementary education? UNESCO recently provided an answer. Since the end of World War II heroic efforts have been made to teach the whole world how to read. The population explosion has largely stultified these efforts. The absolute number of illiterates is greater now than at any time.

The contraceptive revolution which, thanks to advancing science and technology, has made it possible for the highly developed societies of the West to offset the consequences of death control by a planned control of births, has had as yet no effect upon the family life of people in underdeveloped countries. This is not surprising. Death control, as I have already remarked, is easy, cheap, and can be carried out by a small force of technicians. Birth control, on the other hand, is rather expensive, involves the whole adult population, and demands of those who practice it a good deal of forethought and directed willpower. To persuade hundreds of millions of men and women to abandon their tradition-hallowed views of sexual morality, then to distribute and teach them to make use of contraceptive devices or fertility-controlling drugs—this is a huge and difficult task, so huge and so difficult that it seems very unlikely that it can be successfully carried out, within a sufficiently short space of time, in any of the countries where control of the birth rate is most urgently needed.

Extreme poverty, when combined with ignorance, breeds that lack of desire for better things which has been called "wantlessness"—the resigned acceptance of a subhuman lot. But extreme poverty, when it is combined with the knowledge that some societies are affluent, breeds envious desires and the expectation that these desires must of necessity, and very soon, be satisfied. By means of the mass media (those easily exportable products of advancing science and technology) some knowledge of what life is like in affluent societies has been widely disseminated throughout the world's underdeveloped regions. But, alas, the science and technology which have given the industrial West its cars, refrigerators, and contraceptives have given the people of Asia, Africa, and Latin America only movies and radio broadcasts, which they are too simple-minded to be able to criticize, together with a population explosion, which they are still too poor and too tradition-bound to be able to control by deliberate family planning.

In the context of a 3, or even of a mere 2 percent annual increase in numbers, high expectations are foredoomed to disappointment. From disappointment, through resentful frustration, to widespread social unrest the road is short. Shorter still is the road from social unrest, through chaos, to dictatorship, possibly of the Communist party, more probably of generals and colonels. It would seem, then, that for two-thirds of the human race now suffering from the consequences of uncontrolled breeding in a context of industrial backwardness, poverty, and illiteracy, the prospects for democracy, during the next ten or twenty years, are very poor.

A rapid and accelerating population increase that will nullify the best efforts of underdeveloped societies to better their lot and will keep two-thirds of the human race in a condition of misery in anarchy or of misery under dictatorship, and the intensive preparations for a new kind of war that, if it breaks out, may bring irretrievable ruin to the one-third of the human race now living prosperously in highly industrialized societies—these are the two main threats to democracy now confronting us. Can these threats be eliminated? Or, if not eliminated, at least reduced?

My own view is that only by shifting our collective attention from the merely political to the basic biological aspects of the human situation can we hope to mitigate and shorten the time of troubles into which, it would seem, we are now moving. We cannot do without politics; but we can no longer afford to indulge in bad, unrealistic politics. To work for the survival of the species as a whole and for the actualization in the greatest possible number of individual men and women of their potentialities for good will, intelligence, and creativity—this, in the world of today, is good, realistic politics. To cultivate the religion of idolatrous nationalism, to subordinate the interests of the species and its individual members to the interests of a single national state and its ruling minority—in the context of the population explosion, missiles, and atomic warheads, this is bad and thoroughly unrealistic politics. Unfortunately, it is to bad and unrealistic politics that our rulers are now committed.

Ecology is the science of the mutual relations of organisms with their environment and with one another. Only when we get it into our collective head that the basic problem confronting twentieth-century man is an ecological problem will our politics improve and become realistic. How does the human race propose to survive and, if possible, improve the lot and the intrinsic quality of its individual members? Do we propose to live on this planet in symbiotic harmony with our environment? Or, preferring to be wantonly stupid, shall we choose to live like murderous and suicidal parasites that kill their host and so destroy themselves?

Committing that sin of overweening bumptiousness, which the Greeks called *hubris,* we behave as though we were not members of earth's ecological community, as though we were privileged and, in some sort, supernatural beings and could throw our weight around like gods. But in fact we are, among other things, animals—emergent parts of the natural order. If our politicians were realists, they would think rather less about missiles and the problem of landing a couple of astronauts on the moon, rather more about hunger and moral squalor and the problem of enabling three billion men, women, and children, who will soon be six billions, to lead a tolerably human existence without, in the process, ruining and befouling their planetary environment.

Animals have no souls; therefore, according to the most authoritative Christian theologians, they may be treated as though they were things. The truth, as we are now beginning to realize, is that even things ought not to be treated as *mere* things. They should be treated as though they were parts of a vast living organism. "Do as you would be done by." The Golden Rule applies to our dealings with nature no less than to our dealings with our fellow-men. If we hope to be well treated by nature, we must stop talking about "mere things" and start treating our planet with intelligence and consideration.

Power politics in the context of nationalism raises problems that, except by war, are practically insoluble. The problems of ecology, on the other hand, admit of a rational solution and can be tackled without the arousal of those violent passions always associated with dogmatic ideology and nationalistic idolatry. There may be arguments about the best way of raising wheat in a cold climate or of re-afforesting a denuded mountain. But such arguments never lead to organized slaughter. Organized slaughter is the result of arguments about such questions as the following: Which is the best nation? The best religion? The best political theory? The best form of government? Why are other people so stupid and wicked? Why can't they see how good and intelligent *we* are? Why do they resist our beneficent efforts to bring them under our control and make them like ourselves?

To questions of this kind the final answer has always been war. "War," said Clausewitz, "is not merely a political act, but also a political instrument, a continuation of political relationships, a carrying out of the same by other means." This was true enough in the eighteen thirties, when Clausewitz published his famous treatise; and it continued to be true until 1945. Now, pretty obviously, nuclear weapons, long-range rockets, nerve gases, bacterial aerosols, and the "Laser" (that highly promising, latest addition to the world's military arsenals) have given the lie to Clausewitz. All-out war with modern weapons is no longer a continuation of previous policy; it is a complete and irreversible break with previous policy.

Power politics, nationalism, and dogmatic ideology are luxuries that the human race can no longer afford. Nor, as a species, can we afford the luxury of ignoring man's ecological situation. By shifting our attention from the now completely irrelevant and anachronistic politics of nationalism and military power to the problems of the human species and the still inchoate politics of human ecology we shall be killing two birds with one stone—reducing the threat of sudden destruction by scientific war and at the same time reducing the threat of more gradual biological disaster.

The beginnings of ecological politics are to be found in the special services of the United Nations Organization. UNESCO, the Food and Agriculture Organization, the World Health Organization, the various Technical Aid Services—all these are, partially or completely, concerned with the ecological problems of the human species. In a world where political problems are thought of and worked upon within a frame of reference whose coordinates are nationalism and military power, these ecology-oriented organizations are regarded as peripheral. If the problems of humanity could be thought about and acted upon within a frame of reference that has survival for the species, the well-being of individuals, and the actualization of man's desirable potentialities as its coordinates, these peripheral organizations would become central. The subordinate politics of survival, happiness, and personal fulfillment would take the place now occupied by the politics of power, ideology, nationalistic idolatry, and unrelieved misery.

In the process of reaching this kind of politics we shall find, no doubt, that we have done something, in President Wilson's prematurely optimistic words, "to make the world safe for democracy."

BARRY COMMONER

The Closing Circle — Nature, Man and Technology

Barry Commoner is Chairman of the Department of Botany at Washington University, St. Louis, Missouri. He has written and conducted extensive research in the field of environmental pollution and destruction. The following selection is from his profoundly disturbing book, *The Closing Circle* (1971). Here Commoner, long-time ecological warrior, shows that both the causes and cures of the environmental crisis are complex and require a thorough revaluation of our goals and institutions. We must re-examine our basic economic values because the huge environmental debt we have built is coming due.

The environment has just been rediscovered by the people who live in it. In the United States the event was celebrated in April 1970, during Earth Week. It was a sudden, noisy awakening. School children cleaned up rubbish; college students organized huge demonstrations; determined citizens recaptured the streets from the automobile, at least for a day. Everyone seemed to be aroused to the environmental danger and eager to do something about it.

They were offered lots of advice. Almost every writer, almost every speaker, on the college campuses, in the streets and on television and radio broadcasts, was ready to fix the blame and pronounce a cure for the environmental crisis.

Some regarded the environmental issue as politically innocuous:

Ecology has become the political substitute for the word "motherhood."—
Jesse Unruh, Democratic Leader of the State of California Assembly

But the FBI took it more seriously:

On April 22, 1970, representatives of the FBI observed about two hundred persons on the Playing Fields shortly after 1:30 p.m. They were joined a few minutes later by a contingent of George Washington University students who arrived chanting "Save Our Earth." . . . A sign was noted which read "God Is Not Dead; He Is Polluted on Earth." . . . Shortly after 8:00 p.m. Senator Edmund Muskie (D), Maine, arrived and gave a short anti-pollution speech. Senator Muskie was followed by journalist I.F. Stone, who spoke for twenty minutes on the themes of anti-pollution, anti-military, and anti-administration.—FBI report entered into Congressional Record by Senator Muskie on April 14, 1971

Some blamed pollution on the rising population:

The pollution problem is a consequence of population. It did not much matter how a lonely American frontiersman disposed of his waste. . . . But as population became denser, the natural chemical and biological recycling processes became overloaded. . . . Freedom to breed will bring ruin to all.—Garrett Hardin, biologist

The causal chain of the deterioration [of the environment] is easily followed to its source. Too many cars, too many factories, too much detergent, too much pesticide, multiplying contrails, inadequate sewage treatment plants, too little water, too much carbon dioxide—all can be traced easily to *too many people.*—Paul R. Ehrlich, biologist

Some blamed affluence:

The affluent society has become an effluent society. The 6 percent of the world's population in the United States produces 70 percent or more of the world's solid wastes.—Walter S. Howard, biologist

And praised poverty:

Blessed be the starving blacks of Mississippi with their outdoor privies, for they are ecologically sound, and they shall inherit a nation.—Wayne H. Davis, biologist

But not without rebuttal from the poor:

You must not embark on programs to curb economic growth without placing a priority on maintaining income, so that the poorest people won't simply be further depressed in their condition but will have a share, and be able to live decently.—George Wiley, chemist and chairman, National Welfare Rights Organization

And encouragement from industry:

It is not industry *per se*, but the demands of the public. And the public's demands are increasing at a geometric rate, because of the increasing standard of living and the increasing growth of population. . . . If we can convince the national and local leaders in the environmental crusade of this basic logic, that population causes pollution, then we can help them focus their attention on the major aspect of the problem.—Sherman R. Knapp, chairman of the board, Northeast Utilities

Some blamed man's innate aggressiveness:

The first problem, then, is people. . . . The second problem, a most fundamental one, lies within us—our basic aggressions. . . . As Anthony Storr has said: "The sombre fact is that we are the cruelest and most ruthless species that has ever walked the earth."—William Roth, director, Pacific Life Assurance Company

While others blamed what man had learned:

People are afraid of their humanity because systematically they have been taught to become inhuman. . . . They have no understanding of what it is to love nature. And so our airs are being polluted, our rivers are being poisoned, and our land is being cut up.—Arturo Sandoval, student, Environmental Action

A minister blamed profits:

Environmental rape is a fact of our national life only because it is more profitable than responsible stewardship of earth's limited resources.—Channing E. Phillips, Congregationalist minister

While a historian blamed religion:

Christianity bears a huge burden of guilt. . . . We shall continue to have a worsening ecologic crisis until we reject the Christian axiom that nature has no reason for existence save to serve man.—Lynn White, historian

A politician blamed technology:

A runaway technology, whose only law is profit, has for years poisoned our air, ravaged our soil, stripped our forests bare, and corrupted our water resources. —Vance Hartke, senator from Indiana

While an environmentalist blamed politicians:

There is a peculiar paralysis in our political branches of government, which are primarily responsible for legislating and executing the policies environmentalists are urging. . . . Industries who profit by the rape of our environment see to it that legislators friendly to their attitudes are elected, and that bureaucrats of similar attitude are appointed.—Roderick A. Cameron, of the Environmental Defense Fund

Some blamed capitalism:

Yes, it's official—the conspiracy against pollution. And we have a simple program—arrest Agnew and smash capitalism. We make only one exception to our pollution stand—everyone should light up a joint and get stoned. . . . We say to Agnew country that Earth Day is for the sons and daughters of the American Revolution who are going to tear this capitalism down and set us free.—Rennie Davis, a member of the "Chicago Seven"

While capitalists counterattacked:

The point I am trying to make is that we are solving most of our problems . . . that conditions are getting better not worse . . . that American industry is spending over three billion dollars a year to clean up the environment and additional billions to develop products that will *keep* it clean . . . and that the real danger is *not* from the free-enterprise Establishment that has made ours the most prosperous, most powerful and most charitable nation on earth. No, the danger today resides in the Disaster Lobby—those crepe-hangers who, for personal gain or out of sheer ignorance, are undermining the American system and threatening the lives and fortunes of the American people. Some people have let the gloom-mongers scare them beyond rational response with talk about atomic annihilation. . . . Since World War II over one *billion* human beings who worried about A-bombs and H-bombs died of other causes. They worried for nothing.—Thomas R. Shepard, Jr., publisher, *Look* Magazine

And one keen observer blamed everyone:

We have met the enemy and he is us.—Pogo

Earth Week and the accompanying outburst of publicity, preaching, and prognostication surprised most people, including those of us who had worked for years to generate public recognition of the environmental crisis. What surprised me most were the numerous, confident explanations of the cause and cure of the crisis. For having spent some years in the effort simply to detect and describe the growing list of environmental problems—radioactive fallout, air and water pollution, the deterioration of the soil—and in tracing some of their links to social and political processes, the identification of a single cause and cure seemed a rather bold step. During Earth Week, I discovered that such reticence was far behind the times.

After the excitement of Earth Week, I tried to find some meaning in the welter of contradictory advice that it produced. It seemed to me that the confusion of Earth Week was a sign that the situation was so complex and ambiguous that people could read into it whatever conclusion their own beliefs—about human nature, economics, and politics—suggested. Like a Rorschach ink blot, Earth Week mirrored personal convictions more than objective knowledge.

Earth Week convinced me of the urgency of a deeper public understanding of the origins of the environmental crisis and its possible cures. That is what this book is about. It is an effort to find out what the environmental crisis *means.*

Such an understanding must begin at the source of life itself: the earth's thin skin of air, water, and soil, and the radiant solar fire that bathes it. Here, several billion years ago, life appeared and was nourished by the earth's substance. As it grew, life evolved, its old forms transforming the earth's skin and new ones adapting to these changes. Living things multiplied in number, variety, and habitat until they formed a global network, becoming deftly enmeshed in the surroundings they had themselves created. This is the *ecosphere,* the home that life has built for itself on the planet's outer surface.

Any living thing that hopes to live on the earth must fit into the ecosphere or perish. The environmental crisis is a sign that the finely sculptured fit between life and its surroundings has begun to corrode. As the links between one living thing and another, and between all of them and their surroundings, begin to break down, the dynamic interactions that sustain the whole have begun to falter and, in some places, stop.

Why, after millions of years of harmonious co-existence, have the relationships between living things and their earthly surroundings begun to collapse? Where did the fabric of the ecosphere begin to unravel? How far will the process go? How can we stop it and restore the broken links?

Understanding the ecosphere comes hard because, to the modern mind, it is a curiously foreign place. We have become accustomed to think of separate, singular events, each dependent upon a unique, singular cause. But in the ecosphere every effect is also a cause: an animal's waste becomes food for soil bacteria; what bacteria excrete nourishes plants; animals eat the plants. Such ecological cycles are hard to fit into human experience in the age of technology, where machine A always yields product B, and product B, once used, is cast away, having no further meaning for the machine, the product, or the user.

Here is the first great fault in the life of man in the ecosphere. We have broken out of the circle of life, converting its endless cycles into man-made, linear events: oil is taken from the ground, distilled into fuel, burned in an engine, converted thereby into noxious fumes, which are emitted into the air. At the end of the line is smog. Other man-made breaks in the ecosphere's cycles spew out toxic chemicals, sewage, heaps of rubbish—testimony to our power to tear the ecological fabric that has, for millions of years, sustained the planet's life.

Suddenly we have discovered what we should have known long before: that the ecosphere sustains people and everything that they do; that anything that fails to fit into the ecosphere is a threat to its finely balanced cycles; that wastes are not only unpleasant, not only toxic, but, more meaningfully, evidence that the ecosphere is being driven towards collapse.

If we are to survive, we must understand *why* this collapse now threatens. Here, the issues become far more complex than even the ecosphere. Our assaults on the ecosystem are so powerful, so numerous, so finely interconnected, that although the damage they do is clear, it is very difficult to discover how it was done. By which weapon? In whose hand? Are we driving the ecosphere to destruction simply by our growing numbers? By our greedy accumulation of wealth? Or are the machines which we have built to gain this wealth—the magnificent technology that now feeds us out of neat packages, that clothes us in man-made fibers, that surrounds us with new chemical creations—at fault?

This book is concerned with these questions. It begins with the ecosphere, the setting in which civilization has done its great—and terrible—deeds. Then it moves to a description of some of the damage we have done to the ecosphere—to the air, the water, the soil. However, by now such horror stories of environmental destruction are familiar, even tiresome. Much less clear is what we need to learn from them, and so I have chosen less to shed tears for our past mistakes than to try to understand them. Most of this book is an effort to discover which human acts have broken the circle of life, and why. I trace the environmental crisis from its overt manifestations in the ecosphere to the ecological stresses which they reflect, to the faults in productive technology—and in its scientific background—that generate these stresses, and finally to the economic, social, and political forces which have driven us down this self-destructive course. All this in the hope— and expectation—that once we understand the origins of the environmental crisis, we can begin to manage the huge undertaking of surviving it. . . .

I have been concerned with the links between the environmental crisis and the social systems of which it is a part. The book shows, I believe, that the logic of ecology sheds considerable light on many of the troubles which afflict the earth and its inhabitants. An understanding of the environmental crisis illuminates the need for social changes which contain, in their broader sweep, the solution of the environmental crisis as well.

But there is a sharp contrast between the logic of ecology and the state of the real world in which environmental problems are embedded. Despite the constant reference to palpable, everyday life experiences—foul air, polluted water, and rubbish heaps—there is an air of unreality about the environmental crisis. The complex chemistry of smog and fertilizers and their even more elaborate connections to economic, social, and political problems are concepts that deal with real features of modern life, but they remain *concepts*. What is real in our lives and, in contrast to the reasonable logic of ecology, chaotic and intractable, is the apparently hopeless inertia of the economic and political system; its fantastic agility in sliding away from the basic issues which logic reveals; the selfish maneuvering of those in power, and their willingness to use, often unwittingly, and sometimes cynically, even environmental deterioration as a step toward more political power; the frustration of the individual citizen confronted by this power and evasion; the confusion that we all feel in seeking a way out of the environmental morass. To bring environmental logic into contact with the real world we need to relate it to the over-all social, political, and economic forces that govern both our daily lives and the course of history.

We live in a time that is dominated by enormous technical power and extreme human need. The power is painfully self-evident in the megawattage of power plants, and in the megatonnage of nuclear bombs. The human need is evident in the sheer numbers of people now and soon to be living, in the deterioration of their habitat, the earth, and in the tragic world-wide epidemic of hunger and want. The gap between brute power and human need continues to grow, as the power fattens on the same faulty technology that intensifies the need.

Everywhere in the world there is evidence of a deepseated failure in the effort to use the competence, the wealth, the power at human disposal for the maximum good of human beings. The environmental crisis is a major example of this failure. For we are in an environmental crisis because the means by which we use the ecosphere to produce wealth are destructive of the ecosphere itself. The present system of production is self-destructive; the present course of human civilization is suicidal.

The environmental crisis is somber evidence of an insidious fraud hidden in the vaunted productivity and wealth of modern, technology-based society. This wealth has been gained by rapid short-term exploitation of the environmental system, but it has blindly accumulated a debt to nature (in the form of environmental destruction in developed countries and of population pressure in developing ones)—a debt so large and so pervasive that in the next generation it may, if unpaid, wipe out most of the wealth it has gained us. In effect, the account books of modern society are drastically out of balance, so that, largely unconsciously, a huge fraud has been perpetrated on the people of the world. The rapidly worsening course of environmental pollution is a warning that the bubble is about to burst, that the demand to pay the global debt may find the world bankrupt.

This does *not* necessarily mean that to survive the environmental crisis, the people of industrialized nations will need to give up their "affluent" way of life. For as shown earlier, this "affluence," as judged by conventional measures— such as GNP, power consumption, and production of metals—is itself an illusion. To a considerable extent it reflects ecologically faulty, socially wasteful types of production rather than the actual welfare of individual human beings. There- fore, the needed productive reforms can be carried out without seriously reducing the present level of *useful* goods available to the individual; and, at the same time, by controlling pollution the quality of life can be improved significantly.

There are, however, certain luxuries which the environmental crisis, and the approaching bankruptcy that it signifies, will, I believe, force us to give up. These are the *political* luxuries which have so long been enjoyed by those who can benefit from them: the luxury of allowing the wealth of the nation to serve preferentially the interests of so few of its citizens; of failing fully to inform citizens of what they need to know in order to exercise their right of political governance; of condemning as anathema any suggestion which reexamines basic economic values; of burying the issues revealed by logic in a morass of self- serving propaganda.

To resolve the environmental crisis, we shall need to forego, at last, the luxury of tolerating poverty, racial discrimination, and war. In our unwitting march toward ecological suicide we have run out of options. Now that the bill for the environmental debt has been presented, our options have become reduced to two: either the rational, social organization of the use and distribution of the earth's resources, or a new barbarism.

This iron logic has recently been made explicit by one of the most insistent proponents of population control, Garrett Hardin. Over recent years he has expounded on the "tragedy of the commons"—the view that the world ecosystem is like a common pasture where each individual, guided by a desire for personal gain, increases his herd until the pasture is ruined for all. Until recently, Hardin drew two rather general conclusions from this analogy: first, that "freedom in a commons brings ruin to all," and second, that the freedom which must be con- strained if ruin is to be avoided is not the derivation of private gain from a social good (the commons), but rather "the freedom to breed."

Hardin's logic is clear, and follows the course outlined earlier: if we accept as unchangeable the present governance of a social good (the commons, or the ecosphere) by private need, then survival requires the immediate, drastic limita- tion of population. Very recently, Hardin has carried this course of reasoning to its logical conclusion; in an editorial in *Science,* he asserts:

Every day we [i.e., Americans] are a smaller minority. We are increasing at only one per cent a year; the rest of the world increases twice as fast. By the year 2000, one person in twenty-four will be an American; in one hundred years only one in forty-six. . . . If the world is one great commons, in which all food is shared equally, then we are lost. Those who breed faster will replace the rest. . . .

In the absence of breeding control a policy of "one mouth one meal" ultimately produces one totally miserable world. In a less than perfect world, the allocation of rights based on territory must be defended if a ruinous breeding race is to be avoided. It is unlikely that civilization and dignity can survive everywhere; but better in a few places than in none. Fortunate minorities must act as the trustees of a civilization that is threatened by uninformed good intentions.

Here, only faintly masked, is barbarism. It denies the equal right of all the human inhabitants of the earth to a humane life. It would condemn most of the people of the world to the material level of the barbarian, and the rest, the "fortunate minorities," to the moral level of the barbarian. Neither within Hardin's tiny enclaves of "civilization," nor in the larger world around them, would anything that we seek to preserve—the dignity and the humaneness of man, the grace of civilization—survive.

In the narrow options that are possible in a world gripped by environmental crisis, there is no apparent alternative between barbarism and the acceptance of the economic consequence of the ecological imperative—that the social, global nature of the ecosphere must determine a corresponding organization of the productive enterprises that depend on it.

One of the common responses to a recitation of the world's environmental ills is a deep pessimism, which is perhaps the natural aftermath to the shock of recognizing that the vaunted "progress" of modern civilization is only a thin cloak for global catastrophe. I am convinced, however, that once we pass beyond the mere awareness of impending disaster and begin to understand *why* we have come to the present predicament, and where the alternative paths ahead can lead, there is reason to find in the very depths of the environmental crisis itself a source of optimism.

There is, for example, cause for optimism in the very complexity of the issues generated by the environmental crisis; once the links between the separate parts of the problem are perceived, it becomes possible to see new means of solving the whole. Thus, confronted separately, the need of developing nations for new productive enterprises, and the need of industrialized countries to reorganize theirs along ecologically sound lines, may seem hopelessly difficult. However, when the link between the two—the ecological significance of the introduction of synthetic substitutes for natural products—is recognized, ways of solving both can be seen. In the same way, we despair over releasing the grip of the United States on so much of the world's resources until it becomes clear how much of this "affluence" stresses the environment rather than contributes to human welfare. Then the very magnitude of the present United States share of the world's resources is a source of hope—for its reduction through ecological reform can then have a large and favorable impact on the desperate needs of the developing nations.

I find another source of optimism in the very nature of the environmental crisis. It is not the product of man's *biological* capabilities, which could not change in time to save us, but of his *social* actions—which are subject to much more rapid change. Since the environmental crisis is the result of the social mismanagement of the world's resources, then it can be resolved and man can survive in a humane condition when the social organization of man is brought into harmony with the ecosphere.

Here we can learn a basic lesson from nature: that nothing can survive on the planet unless it is a cooperative part of a larger, global whole. Life itself learned that lesson on the primitive earth. For it will be recalled that the earth's first living things, like modern man, consumed their nutritive base as they grew, converting the geochemical store of organic matter into wastes which could no longer serve their needs. Life, as it first appeared on the earth, was embarked on a linear, self-destructive course.

What saved life from extinction was the invention, in the course of evolution, of a new life-form which reconverted the waste of the primitive organisms into fresh, organic matter. The first photosynthetic organisms transformed the rapacious, linear course of life into the earth's first great ecological cycle. By closing the circle, they achieved what no living organism, alone, can accomplish— survival.

Human beings have broken out of the circle of life, driven not by biological need, but by the social organization which they have devised to "conquer" nature: means of gaining wealth that are governed by requirements conflicting with those which govern nature. The end result is the environmental crisis, a crisis of survival. Once more, to survive, we must close the circle. We must learn how to restore to nature the wealth that we borrow from it.

In our progress-minded society, anyone who presumes to explain a serious problem is expected to offer to solve it as well. But none of us—singly or sitting in committee—can possibly blueprint a specific "plan" for resolving the environmental crisis. To pretend otherwise is only to evade the real meaning of the environmental crisis: that the world is being carried to the brink of ecological disaster not by a singular fault, which some clever scheme can correct, but by the phalanx of powerful economic, political, and social forces that constitute the march of history. Anyone who proposes to cure the environmental crisis undertakes thereby to change the course of history.

But this is a competence reserved to history itself, for sweeping social change can be designed only in the workshop of rational, informed, collective social action. That we must act is now clear. The question which we face is how. . . .

HENRY L. DIAMOND

Our Lagging Institutions

Henry L. Diamond is Vice-President of the American Conservation Association, New York. He is also an attorney and has served as an advisor to the Presidential Committee on Environmental Quality. In this selection, the title of which is somewhat misleading, Diamond sees our existing institutions as slow, but in the long run responsive to public opinion. They are, in fact, responding, he believes, to a newly enlightened electorate. Some shoring-up is needed, but the outlook is bright.

We are in the midst of an environmental revolution. The people of this country are really sick of the degradation of their environment, and they are demanding that their leaders and their institutions take action.

Not surprisingly, our broadly based institutions are not set up to respond as quickly and as aggressively as the people want. In the past, environmental concern has been a very minor theme of our society. It was in the province of the little old ladies in sneakers and the dotty brothers-in-law in tweeds. Our institutions simply did not have to deal with this issue.

Now they must.

Let me list the state of our institutions with regard to their concern for environment quality:

Government. Government at all levels, but particularly the federal, has been literally unable to pull itself together to face the rising environmental aspirations of the American people. This year the President by Executive Order, and the Congress by legislation, are trying new ways. There are problems, but a start is being made.

The Courts. Environment has not been a subject of litigation, but this is rapidly changing. The law may be the new frontier of environmental action. Half a dozen organizations are springing up to help people take their environmental grievances to court. The preliminary indications are that the courts are responsive, and we may be seeing a whole new body of law in the making.

The Press. For many years reporting the environment consisted chiefly of stories on how the trouts were running. But now good newspapers are springing top reporters for environment and printing their stuff. And if *Time* now has an Environment section, can *Newsweek* be far behind?

The Church. The Church has approached the environment chiefly from the point of the use of leisure time. Sloth being one of the first-order sins, the Church has worried about good, wholesome outdoor recreation for people when they are not working. Increasingly, however, the Church is becoming concerned with

Henry L. Diamond, "Our Lagging Institutions," from *No Deposit-No Return,* edited by Huey D. Johnson, 1970, Addison-Wesley, Reading, Mass. Pp. 172–174.

environment and "Thou shalt not throw crap untreated into the river" may become a pretty serious sort of commandment.

Education. A number of school systems have begun to teach environmental awareness with programs financed by the Education Act of 1965. New college courses, departments, and programs seem to be proliferating.

Unions. Some larger and more sophisticated unions have established departments to worry about the environment. Understandably the union position on any given issue is likely to be determined by whether jobs are created or not, but there is a trend toward environmental awareness. The UAW's redoubtable Olga Madar is an excellent example.

Business. There is a growing awareness of the importance of environmental quality in the corporate board rooms. Business is finding that the public reaction to what they want to do to the environment may become an important factor. Business is learning that what was once regarded as only a lunatic fringe is now a formidable opponent armed with lawyers, public relations men, and even voting stock.

Foundations. Five years ago one could name on the fingers of one hand the foundations that cared anything about the environment. Since then, Ford has come onto the scene in a big way, and scores of smaller foundations have an awakened interest.

I believe that this trend toward greater involvement in environmental affairs by our major institutions will continue. I am sanguine for a very basic reason. Years ago Mr. Dooley noted that the Supreme Court followed the election returns pretty well. Well, the same is true of our institutions. The bureaucrats, the businessmen, the churchmen, the foundation people, the publishers, all those who make policy for our institutions, keep their fingers pretty close to the public pulse. They watch not only the election returns but the polls, the buying patterns, the letters to the editors, and the hell-raisers.

And anyone who's doing any watching at all today should be able to discern that environment is "in." It has taken its place as a new, major concern of the American people.

In my view the real institutional lag—the most crucial one—is among the environmental organizations themselves. These institutions should be in the first wave of the environmental revolution. Public pressure is the heart which makes the larger institutions move. The environmental organizations must carefully ignite, fan, and direct the brush fires into white hot issues. But the existing organizations have been caught unaware and unprepared.

In the first place, there were no real environmental organizations. There were a number of highly specialized groups in the fields of conservation, outdoor recreation, natural resources, and city betterment, but none really had the scope of the entire environment. Traditionally these organizations have appealed to a limited membership. They existed to provide more birds as shotgun fodder— not to build a broad environmental coalition. Secondly, these organizations commanded relatively few resources—money and staff have been spread pitifully thin.

Thirdly, the existing organizations are not attuned to the new scene. The people who are generating environmental revolution are a whole new constituency— young, involved, politically committed, scientifically aware, and anxious for action. Nothing will turn them off faster than a dusty, old organization worried more about book sales and bird lists than clobbering the bad guys.

Some of these organizations are trying to broaden their base—to catch up in this environmental lag. Some are coming in from the wilderness, and some are coming out from the city, to cover the full range of the environment.

It may be, however, that we need a reshuffling of our environmental organizations. The most effective and most alive organizations are the new ones springing up at local levels. Quite often people have come together over a single issue and stayed together to fight other battles. These *ad hoc* groups have a vitality and a drive which has been lacking. They often are made up of politically involved people who know how to use campaign techniques. They are where the action is.

At present, I do not see any national environmental organizations able to serve these new action people across the country. It is these new, zealous converts who will keep environmental revolution going, and they will fire up the major institutions.

L. W. LANE, JR.

An Environmental Ethic

L.W. Lane, Jr., is a California publisher and politician. He has been active in the struggle against the unlimited use of DDT. "An Environmental Ethic" seeks what might be called a "spiritual" solution to the environmental problem. Lane advocates the application of Christian charity, a sort of environmental altruism, which will not necessitate any basic changes in our economic arrangements. He espouses an attitude which he believes will supplant pecuniary preoccupations with a concern for a long-term investment in our environment.

Laws and ethics on how people behave toward one another, and as groups in a socio-political-economic environment, have usually come about from absolute necessity and very often a crisis. Rarely do these codes of behavior rise to a level of wide acceptance and enforceable laws out of the simple wisdom and great foresight of a few people. The pot only begins to boil when the heat builds up under it. History will show that many of the laws which we all take for granted first saw the light of day in the writings and proclamations of crusaders and zealots, often regarded as radicals and crackpots, who were truly playing the role

L.W. Lane, Jr., "An Environmental Ethic," from *No Deposit-No Return,* edited by Huey D. Johnson, 1970, Addison-Wesley, Reading, Mass. Pp. 223–229.

of David fighting the Goliath of public apathy and ignorance. With the pressures
for attention and dollars in this man-on-the-moon period of our history, a great
many of the 202,000,000 people in this nation will have to start hurting before
the heat gets intense enough to generate a fire.

It is not my purpose to give elaborate detail to substantiate the environmental
problems we live with and will eventually face in the future. I do wish to establish
the premise that we are now facing scattered crises and that, as we look into the
crystal ball, recognizing the rate of deterioration and the generally increasing
trends causing the problems, they can only multiply. We must make a national,
if not international, effort to develop a code of ethics and a body of law to
govern the relationships within our environment, which are just as important to
our society, system of government, and life itself as any others ever adopted.

We must develop a national "Environmental Ethic" to inspire and guide us
down this very difficult road. We must, as a people and as a nation, re-identify
our national values and goals to recognize perhaps the most basic and worldwide
problem we face.

Our country and the free enterprise system that our constitution and democratic processes have made possible—and generally encouraged and protected by
the laws of our land—have evolved from a demand for personal freedom and
economic self-interest. The crisis facing our forefathers was to gain freedom on
many fronts political, economic, and religious—in order to reap the rewards of
the good life in the New World. The Constitution was created to set up a work-
able democratic system to protect those freedoms. The Preamble of the Constitution was very precise: justice, domestic tranquility, general welfare, common
defense, and liberty were the touchstones of that day.

Survival and enjoyment of life per se, environment itself, was not in jeopardy
in 1776. Environment was certainly not facing a crisis either. There was a great
abundance of natural resources and a low consumption rate that gave no threat
to the world of nature. Native plant and animal life seemed almost limitless;
there was little to pollute the air and water. When the good land ran out and
neighbors got a little too close, all that was needed was to cross a mountain
range to the west to reach another valley, and there was more land with good
water to settle for the asking. But now we have migrated as far west as we can,
and the state by the Pacific Ocean has more people than any other state in
the Union.

To launch a national program of action, we must have, on a nationwide scale,
initial acceptance by the responsible opinion leaders in the fields of ecology,
industry, communication, education, church, conservation, and government at
all levels. We must evolve, accept, and practice an "Environmental Ethic" that
provides a national code of standards similar to the Golden Rule and the Ten
Commandments, to give us guideposts for our personal and moral conduct.

The acceptance of this national environmental ethic and the crusade to implement it must eventually involve every citizen of this country—and in fact, will
require the cooperation of countries around the world on both sides of all
curtains, whether whey be made of iron or bamboo. The problem knows no
limits of race, color, creed, faith, income level, political party, or boundary.

Environment is the one great common denominator for all people. More than any economic, political, religious, or ethnic binder, the natural world which embodies the atmosphere that belongs to all of us, the water supply that circulates around the earth, the resources held by collective land which we all dig and drill, the animal and plant life so dependent on these common possessions, and all the social problems we must solve—no one subject can do more than this can to bring us together and unite us in common goals that also solidify the spirit and create a "oneness."

Because laws and ethics tend to limit freedom, they are resisted. The human plea, "Don't fence me in," is strong in every breast. Yet we have been forced to face the problem of preventing jungle war in other areas, to bring order and enforce proper behavior through the Pure Food and Drug Act, Robinson Patman Act, Sherman Anti-Trust Laws, Taft-Hartley Law, and many more.

In certain cases the challenge is primarily one of changing attitudes and implementing laws already on the books. This is a substantial and often readily available opportunity to protect our environment. Much painful but definite progress in civil rights is coming about in this way. Generally speaking, existing laws give planning commissions and all government bodies and individuals dealing with environmental matters far more authority to take firm action than is often exercised. In my experience in either participating in or witnessing all levels of governmental action in this field, the tough decision is often avoided, not for lack of laws, regulations, and ordinances, but for lack of sufficient information, an absence of a strong code of ethics, and, sad to say, a lack of guts.

One of man's special traits is to ask thoughtful questions about what he should do or not do. Aristotle, who put the word ethics into the common language, stressed the ethical significance of the fact that all men seek happiness. This rationale was carried further by defining the greatest happiness as coming from the contemplative use of the mind, according to Aristotle. "Peace of mind" is an increasing objective in many environmental situations today.

Perhaps most basic, as we think about the slowly emerging "Environmental Ethic" in this country, is that an obligation and sacrifice of freedom by the individual is inherent in any ethic. Aldo Leopold wrote in his farsighted book of 20 years ago, in his chapter on "The Land Ethic": "An ethic, ecologically, is a limitation of freedom of action in the struggle for existence. As an ethic, philosophically, it is a differentiation of social from anti-social conduct."

Until a few years ago there were few, if any, better illustrations of the concept of this country as a "sweet land of liberty" than man's attitudes and actions related to environment. He could cut, burn, pollute, bulldoze, dredge, fill, and foul up the environment on his plot of ground or subdivision, out of his car or smokestack, or in his well or irrigation district in just about any way he wanted to!

The laws of our land have pegged our national values to the growth and profit of our economic system of free enterprise. Land tends to be considered for zoning and tax purposes in terms of its highest economic value. In fact, our ethics governing land and other resources, including the water on and the air above the land, are strongly influenced by economic self-interest. Stewart Udall once referred

to the U.S. GNP as our Holy Grail. It is true that the measures we apply to
success generally emphasize quantity rather than quality, notwithstanding many
examples that quality can be good business.

In a trickle of examples that are rapidly forming a stream of environmental
success stories, we see more and more evidence of a public demand for correcting
and preventing environmental failures and of industry and government response
to environmental problems.

We are evolving an "Environmental Ethic," or at least decisions and actions that
are encouraging signs of a change in thinking toward environment. It is not
quite the abstract and stuffy word of a few years ago. It is more and more an
"in" word. It is being increasingly accepted for its total value including physical
and mental health, community values, and often values that are economically
sound.

What we must believe in, as a part of our "Environmental Ethic," is a critical
premise that is behind most environmental and ecological values. Any one
individual, or one industry, or one local government is a member of a community
of interdependent parts and must function cooperatively to determine his
destiny.

A city 500 miles down the Mississippi River from another city in a different
state, several days apart by stagecoach or riverboat when they were founded, is a
20th-century neighbor environmentally, just as much as adjacent communities
share their environmental problems of smog, water, traffic, zoning, floods, sewage,
and many more. Environmental anarchy by a homeowner, a business, a local
government, or a country is becoming more and more intolerable.

Because a growing number of Americans feel this in all walks of life, we find a
definite break in the logjam. The greatest progress is coming in one of the
easiest but very critical areas in which to drive home the need for "oneness":
air and water. Because the very cyclical pattern of air and water gives less opportunity for legal proprietary rights—and because air and water are recognized for
their common value to all of us—we are finding some tough laws being passed
and some voluntary efforts that we hope will correct existing problems and
prevent future ones. Some situations unfortunately are perhaps beyond complete
correction ecologically. If there is any silver lining in that cloudy situation, it is
the fact that the Lake Eries, the Hudson Rivers, the Lake Tahoes, and the San
Francisco Bays have fanned the fire of crisis to generate the heat that starts the
pot boiling in public opinion, industry awareness, and government enforcement.

Recently, seven major airlines decided to equip all new planes with smokeless
engines and convert 3000 existing engines. This "voluntary" step came as a
result of the airlines being named as defendants in a suit by the New Jersey State
Department of Health charging excessive pollution of New Jersey's air.

In Connecticut, 150 young schoolgirls waged a hard campaign from scratch
called PYE—"Protect Your Environment." The result was unanimous adoption
by the State Legislature of a bill calling for a master plan survey and an immediate set of protective laws for preserving the coastal and tidal areas from dumping.

In San Mateo County in California, the Regional Planning Committee brought
together all local city jurisdictions to agree on an open space acquisitions

program. A direct result of the educational aspect of this study was the decision to set aside 23,000 acres of the San Francisco watershed in perpetuity for open space in a joint power agreement among the County, the City of San Francisco, and the state and Federal governments.

There is a grass-roots awakening and there is lots of action going on, but we need to move faster; in comparison with the speed at which we should be moving, we are only crawling.

One of the most significant trends in our society is the emerging affluency of the blue collar segment of the population. Increasingly the blue collar worker has a boat or camper, perhaps a second home, and finds himself all to often fishing or camping by a stream or lake that is polluted by the very economy that has given him his good life. Like his union vote, his franchise as a citizen in the privacy of the voting booth is powerful.

While there are some discomforting aspects of the so-called campus revolt, young people are increasingly and very properly becoming involved in environmental issues. While they occasionally tend to be emotional and unrealistic, and tend to forget that their university is often supported by taxes and contributions that are possible only in a free enterprise and democratic system, they are coming up with some good thinking and action.

But the job has only begun—because of the magnitude of the problems and the rapid advance of many of the causes of the problems. Increased population, leisure time from longer vacations, shorter workdays and work weeks, longer active lives, more material goods and waste to dispose of, and a scarcity of virgin valleys to move on to—all of these are on the long-term upswing. There's open land, but we've reached the end of the road in terms of any mass migrations to ease the pinch of civilization. A few can escape, but the problems are of the many and not the few. For the last several years, we have seen a migration backlash, with both Arizona and Nevada receiving their largest number of migrants from California.

In our own company, we have tried to carry a spear and to attempt to practice a code of ethics. If we consider physical health and the welfare of the family and classroom readership of *Sunset* magazine as part of our total environmental world, our never accepting hard liquor advertising, and none for tobacco products for many years, is a dollars-and-cents reflection of how we are expressing our own code in actual business practice. The many categories of advertising not accepted by our publication comprise some 20% of the dollar volume spent in our industry. When we supported the creation of a controversial national park, we anticipated and promptly received large advertising cancellations. As an encouraging measure of progress in the last few years, I honestly feel that these advertising pages would not be canceled today. Hopefully, our ethical philosophy will be understood as we have just discontinued accepting two-wheels-off-the-road motor vehicle advertising and are applying even tough controls on real estate and land development advertising. There is no point kidding yourself—following any ethical course, including environmental, means you don't always pick up the marbles after the game is over. To think otherwise is economic myopia.

Recently we completed our investigation of DDT and several other hydro-

carbons and their role in home gardening. Our August 1969 issue carried an
article on "Blowing the Whistle on DDT." Because it has been our policy to
adhere to similar policies or codes of ethics, for both the editorial and advertising
content, we discontinued accepting DDT advertising effective with the same
issue. We had heretofore carried more of this advertising category for home
gardening use than any other U.S. publication and we were somewhat surprised
to have the decision featured in *The Wall Street Journal* and to find the full
announcement read into the Congressional Record. The decision, we have learned,
has been used to strengthen arguments for tougher controls and legislation on
pesticides in several states. The mail support has been very heavy in support of
the position we took. These reactions only help to emphasize that there is a
swell of support for "get tough" action.

The challenge for the communication media, and for any organization or
individual is not to get trapped into being "all things to all people." The whole
challenge of conservation is to create an orchestra of activity. Each of us has a
part we can play better than some other. Some play the drum and some play the
violin. We feel our role is to find successful accomplishments and give the factual
"how-it-was-done" information to guide and inspire others. We accept the
premise that to recognize good, we by no means have to ignore the bad.

Realistically, the worldwide recognition and strong leadership from a Secretary
and a Department of Environment would be a positive way to achieve many
of the same goals that have been proposed as a goal for a possible Secretary of
Peace. There is no better place for leadership to be exercised to develop a national
"Environmental Ethic" than from the investigation and deliberations of the
President's Council on Environmental Quality and its Citizens' Advisory Committee on Environmental Quality, whose chairman is Mr. Laurance Rockefeller.

In his announcement of the Council, President Nixon quoted a statement made
by a former President that read: "The conservation of our natural resources and
their proper use constitute the fundamental problem which underlies almost
every problem of our national life. . . ."

We must all pray and work together to make the noble words and plans for
action of 1969 witness for more immediate action and long-term results—more
than those earlier strong words of advice given by Teddy Roosevelt in 1907
have received in the 62 years since. Time is running out, just as many of our
resources are drying up—or soon will be. Noble words and occasional accomplishments are not enough. We must wage total war to win the battle to save and
protect the basic physical environmental elements and social environmental
values that sustain life and make it all worthwhile.

If we wait too long to develop and practice a strong "Environmental Ethic," it
might well start off, "Thou shalt not kill—ourselves."

252

E. W. PFEIFFER AND
A. W. WESTING

Land War

E.W. Pfeiffer is Professor of Zoology at the University of Montana. Arthur H. Westing is Professor of Botany at Windham College, Putney, Vermont. These reports on the ecological effects of the "land war" in Indochina, supported by the Scientists' Institute for Public Information and the Journal *Environment,* were prepared on the scene. What we are doing to our small planet is painfully described in this paper. This is not warfare against a cruel enemy, but against the very earth on which we dwell. That we engage in this sort of self-destructive behavior may very well point up the inadequacy of the kinds of solutions offered by the two preceding papers.

Craters

During the Indochina War the United States has dropped more than two times the tonnage of bombs that was dropped in Europe, Asia, and Africa during World War II, most of it in Vietnam, a country about the size of New England or one-half the size of the state of Montana. Rockets, artillery shells, and mines have been exploded on a vast scale in many areas in Vietnam, in addition to explosives dropped from aircraft. This ordnance has been used principally in free-fire zones or special strike zones, which all people except the National Liberation Front and its North Vietnamese allies have supposedly vacated. Data on the extent of the free-fire zones of South Vietnam would permit calculation of the percent of Vietnamese land surface that has been intensively subjected to these weapons. These data are not, however, presently available.

Although few details have been released regarding expenditures or target locations for the various types of munitions, the following summary figures for all of Indochina have been made available by the Department of Defense:

Munitions Used in Indochina War (in Millions of Pounds)

Year	Air munitions	Surface munitions	Total
1965	630	?	630
1966	1,024	1,164	2,188
1967	1,866	2,413	4,278
1968	2,863	3,003	5,866
1969	2,774	2,808	5,583
1970	1,955	2,389	4,344
Total	11,112	11,777	22,889

We do not know what fraction of the 23 billion pounds of munitions expended during these six years was small arms and other ordnance that would not produce craters (nor do we know what the distribution is among South Vietnam, North Vietnam, Cambodia, and Laos). To make some wild assumptions, if half the munitions (by weight) were of the sort that produce craters (bombs, shells, etc.) and if each was a 500-pound bomb, then Indochina's landscape would now be more or less permanently rearranged by more than twenty million craters. Using an estimated average diameter of 30 feet, the holes alone would cover a combined area of about 325,000 acres. Although occasional, scattered craters can be found almost anywhere in rural South Vietnam, we have observed large areas of severe craterization in the provinces of Tay Ninh, Long Khanh, Gia Dinh, Hau Nghia, Binh Duong, Quang Ngai, Quang Tin, and Quang Nam. We have been told about similar areas in Kien Giang, An Xuyen, and Quang Tri. No type of habitat seems to be spared, including forests and swamps, fields and paddies. Many severely craterized areas—such as the so-called free-fire zones, free-bomb zones, or specified strike zones—were formerly inhabited and farmed. Such regions of important military activity as War Zones C and D, the Iron Triangle, the Rung Sat and U Minh Special War Zones, the Demilitarized Zone, and the Ho Chi Minh Trail are among those regions that have been subjected to repeated saturation or pattern bombing.

What is this unprecedented bombardment doing to Vietnam and its people? In order to make a preliminary assessment of the effects of these explosives, Arthur H. Westing and I visited Vietnam in August 1971. In preparation for our trip we had sought information from many sources on effects of bomb craters resulting from military activities, but were unable to find any significant information.

We flew over bombed areas in helicopters and rode in armored personnel carriers to observe at first hand craters from B-52 strikes. We interviewed in the field Vietnamese farmers who were trying to reclaim bombed land, Vietnamese loggers who were operating in bombed and shelled areas, and several Vietnamese and American officials.

In order to judge the magnitude of the problem it is necessary to have some idea of the number of bombs dropped and the amount of territory affected. Earlier studies have presented data which suggest that some 7.5 million craters have been formed as a result of the massive bombardment. Although we estimate that the current figure for South Vietnam is in excess of 10 million, we are currently awaiting Department of Defense data to verify this figure.

The standard weapon of the B-52s is a 500-pound bomb; each B-52 carries 108 five-hundred-pound bombs. Each bomb produces a hole 20 to 50 feet wide and 5 to 20 feet deep, depending on soil conditions. The bombs are usually dropped from over 30,000 feet by the B-52 aircraft and can have sufficient force on impact to penetrate deeply into certain types of soil.

Severely bombed areas observed on our trip included the following land types: heavily cultivated areas of the Mekong Delta, intensively cultivated mountain valleys in the northern region of Vietnam, mangrove forests, evergreen hardwood forests of the flat terraces northwest of Saigon, and evergreen hardwood forests of the precipitous mountain areas in the Da Nang-Quang Ngai area.

Because of the war situation at the time of our visit, we were unable to fly over, even at high altitude, the most intensively bombed regions of South Vietnam which lie in the northwest corner of the country and along the Demilitarized Zone. We were also very disappointed to find that security problems made it very difficult to visit on foot bombed areas in all of the regions that we attempted to study. It is important to note that there are areas of South Vietnam, particularly in the delta region, that do not reveal, at least from the air at 3,000 feet, much evidence of war damage. Large areas, however, have been hit very intensively by several types of ecologically devastating weapons.

What are the effects of the massive bombardments on cultivated areas such as the Mekong Delta? Our observations made both in wet and (on previous visits) dry seasons show that in the delta the B-52 craters and those caused by large artillery shells are permanently filled with water, probably because the craters penetrate the water table. In many areas waters of different colors fill adjacent craters. Some of the waters in the craters are aquamarine while others have a more bluish to greenish tint, and many are simply a muddy brown. These differences in coloration are apparently due to growths of varying types of algae. It is interesting that different growths occurred in contiguous craters.

I was able to visit on foot three such craters in an agricultural area about 30 miles south of My Tho in the heart of the Mekong Delta. The area, near the hamlet of Hoi Son, had been a free-fire zone until fairly recently, but farmers were now being resettled on their land because senior officials considered the region relatively secure. The degree of security became evident: During my stay in the area U.S. aircraft were rocketing and strafing only a few miles away. I interviewed some families who had left the area eleven years ago because of the fighting. They took me to three craters made in 1967. I would estimate that they were caused by 500-pound bombs dropped by fighter bombers. Each crater was about 30 feet in diameter, filled with water and, at the time of my visit, about 5 feet deep in the center, as proven by one of my guides. He waded into the center of the crater where he could just manage to keep his nose above water while standing. The entire immediate vicinity had been a rice paddie; the rice had been replaced by a very tall reed (6 to 8 feet), genus *Phragmites,* which surrounded the craters at a distance of 10 to 20 feet. Growing from the rim of the craters and into the reeds was a species of relatively short grass, genus *Brachiaria.* A taller grass, *Scirpus,* was also prevalent. The whole area was inundated by very shallow water, as it was the middle of the wet season. The farmers were growing seed rice near the craters and were plowing under the reeds and grasses in preparation for planting rice. It was obvious that they could not use the cratered areas for rice cultivation, because the water was much too deep. One solution to the problem is to bring in soil from elsewhere. Although I could not confirm it, one farmer said that the craters I observed yielded exceptionally good fish catches. The fish presumably had moved into the craters during the monsoon flooding. Surrounding the area that had been cultivated in rice were banana, coconut, and jackfruit trees. The jackfruit was dead as a result of herbicides; the coconut trees were destroyed by the bombing, leaving only bare stumps.

In our conversations with these and other farmers who were trying to resettle their fought-over land, it became obvious that their main problem was the presence of unexploded munitions in the areas. The Hoi Son people stated that within the last few weeks three women had been killed and one badly wounded when plows detonated unexploded weapons. We learned that mines in some resettled areas have been cleared, but the problem of locating and neutralizing unexploded ordnance before land is resettled is an urgent one. On several occasions we encountered the fear of unexploded munitions, which probably accounts for a phenomenon we often observed from the air: fields with craters were usually not being cultivated although nearby fields were. One farmer whom we interviewed stated that the people do not like to plow in the bombed areas because the shrapnel in the dirt cuts the buffalos' hoofs, resulting in infection.

According to science spokesmen of the U.S. Agency for International Development (USAID) and the Military Assistance Command, Vietnam (MACV), bomb craters are sometimes used as sources of fresh water for irrigation. In much of the southern Mekong Delta, brackish (salty) water floods cultivated lands at high tides if it is not kept back by dikes. Thus, irrigation is necessary and fresh water in the craters could be useful.

Presumably the permanently water-filled crater areas of the delta region are excellent breeding grounds for certain species of mosquitoes and other carriers (vectors) of disease. Those craters not invaded by predators of mosquito larvae provide conditions for greatly accelerated reproduction of mosquitoes and other vectors. According to MACV-Command Information pamphlet 6-70, February 1970, "malaria has been causing increasing concern in Vietnam. . . . Up until recently it (*Plasmodium falciparum*) only affected regions of I and II Corps but has now spread to other areas throughout the country." We discussed with several scientists the possible relationship between craterization and this increase in malaria, but no studies have yet been made of this problem as far as we could determine. A USAID specialist in public health with headquarters in Saigon stated that the current alarming increase in hemorrhagic (dengue) fever seen in the Vietnamese was *not* related to craterization because the mosquito vector for this disease, *Aëdes aegypti*, lives only in and around houses and would thus not be affected by ecological changes such as craterization. (We do not know of any field research which supports this view.) We flew a mosquito-control spray mission in a C-123 aircraft from which malathion was being sprayed (one-half pound per acre) over and around an Australian military base. There are only two aircraft now carrying out this program, and, as far as we were able to determine, there is no spray program involving treatment of cratered areas.

We observed many craters in isolated mountain valleys near Da Nang. They were in small clusters in mountain rice fields and thus were probably caused either by artillery or fighter-bomber strikes and not by B-52s. In these valleys the craters were generally filled with water as in the delta, but they probably are without water in the dry season and thus cannot be used for fish culture. The paddies that had been cratered were not being cultivated. During our visit we flew over many rice paddies with ponds in the centers almost comparable in size (about ten feet across) to the bomb craters, but these were fish ponds and apparently did not interfere with the cultivation of the rice surrounding the ponds. It is thus unlikely that scattered craters could create changes in soil moisture or other conditions that would make cratered paddies uncultivatable.

We observed from the air large areas of the mangrove swamps of the Rung Sat Zone which had been subjected to very heavy B-52 strikes. These are all permanently water-filled and obviously would make transportation into the area very difficult. This could be of some significance because the mangrove forests have been regularly used as sources of wood for charcoal and for fishing grounds.

We observed many craters at first hand in the Boi Loi woods area. This had been an evergreen hardwood forest on the flat terrace northwest of Saigon. Most mature trees were dead from defoliation (herbicide spraying) but there was a very thick understory of useless broadleafed brush, vines, bamboo, and *Imperata* grass reaching a height of 15 to 20 feet. Craters were very numerous in this area and were scattered at least every 100 feet or so. Each crater was 20 to 30 feet across and 5 to 10 or more feet deep. They were all in a grey podzolic soil (a poor soil often formed in cool, humid climates) with poorly defined horizons (layers). There were many generations of craters. The most recent ones were bare of vegetation but contained a little rain water at the bottom. In the older ones a few sprigs of grass, probably *Imperata,* were sprouting in the center. (We also noted the beginning of plant growth in the center of some of the water-filled craters in the delta.) As the craters age the grass grows radially, covering the bottom, and finally grows up the sides to meet vines growing down from the peripheral vegetation. There is some filling of old craters with soil washed down from the sides, but this is limited because old craters completely covered with grass were still 5 to 10 feet deep. We did not observe any broad-leafed plants invading these holes.

We were able to learn something of the effects of saturation bombing and artillery fire upon forest timber resources through interviews with loggers and saw mill operators and by inspection of damaged logs, mainly in the Ben Cat and Chon Thanh areas. We also interviewed South Vietnam forestry officials about the problems of utilizing bombed forest areas. These officials indicated that loggers do not like to operate in bombed timber because the trees have metal fragments in them which greatly reduce the value of the logs. (One logger estimated that the price of logs containing metal is reduced by 30 percent.) We could understand the reason for the reduced value when we observed piles of saws with teeth ripped out and examined discarded logs from which we dug pieces of metal. In some logs there were dead areas about twelve inches in diameter and six inches deep from which we recovered bomb fragments. We learned that when mature timber is punctured by metals such as steel shards or bullets, entry is provided for disease organisms, probably fungi, which result in dead areas that increase in size as the wound ages. Thus, largely unlike trees in temperate zones, the trees of Vietnam are susceptible to rot when penetrated by metal. This greatly decreases the value of the timber and also weakens trees so that they are much more subject to being blown down. An official of a French rubber plantation told us that he had lost many rubber trees on his plantation because the trees had been weakened by fungous infection following bomb damage and then blown down in one of the frequent violent wind storms that occur in the area. The loggers whom we interviewed said that the craters in the

forest made passage very difficult for trucks and loaders, a situation that necessitated cutting much shorter logs than desirable in such areas. (We saw 90 foot
logs coming out of undamaged forests.)

We were able to observe from a high-flying helicopter the craters caused in a
mountain forest near Da Nang by a B-52 strike about one and one-half years
earlier. The craters were still obvious on the mountainside and along the ridges.
The large burned areas in these forests appeared to be even more significant;
they had apparently resulted from fires started by various types of ordance such
as white phosphorus, napalm, and flares.

We tentatively conclude that those cultivated areas hit heavily with conventional high explosives will be very difficult, if not impossible, to recultivate. They
can perhaps be used as fish-rearing ponds or, in certain situations, as sources of
fresh water for irrigation. They may provide additional breeding areas for insect
vectors of disease. In the forested areas that have not been killed by chemical
defoliation, the bombing has created problems that are probably just as great as
those caused by defoliation. However, the immediate problem of greatest
concern is the vast number of unexploded mines, bombs, rockets, and so forth,
that must be removed if the land is to be resettled. Since the Department of
Defense reports that approximately 1 to 2 percent of our air and ground munitions fail to explode, there are several hundred thousand of these randomly
buried throughout Indochina.

We recommend studies to determine the relationship of water-filled craters to
the spread of certain diseases, and to determine how cratered areas can best be
rehabilitated. We also recommend that greatly expanded operations be initiated
to locate and neutralize unexploded ordnance in agriculturally useful areas.
(*E. W. Pfeiffer*)

Leveling the Jungle

Despite the lavish application of great wealth and superior technology, the
United States has made surprisingly little headway over the years against the
National Liberation Front and its North Vietnamese allies. With the growing
realization that the forest functions as a key ally of guerilla fighters by providing
cover and sanctuary, more and more effort has been directed tward its obliteration. For a number of years reliance was placed primarily on chemical destruction.
This approach reached its peak in 1967, but largely because of pressure exerted
by the scientific community, it now not only has been reduced to a low level
(see *Environment*, July/August 1970, p. 16) but also has been entirely "Viet-
namized." The herbicidal assault has left South Vietnam with a legacy of many
millions of dead, now rotting trees, and with locally debilitated ecosystems. A
second approach that has been employed through the years to make the forest
less hospitable to the other side is a bombing and shelling program of incredible
magnitude. The 23 billion pounds of total munitions expended in Indochina
between 1965 and 1970 alone are more than double those used by us throughout
World War II in all theaters.

In recent years, however, a new technique has emerged. Born about 1965, developing into major proportions in 1968, and growing ever since, a vast program of systematic forest bulldozing now exists. The U.S. Engineer Command in Vietnam is daily putting Hercules and his twelve labors to shame. This report outlines the methods, scope, and magnitude of this "jungle eating" program and speculates on its economic and ecological impacts.

The basic tool of the land-clearing operations in Vietnam is the 20-ton D-7E Caterpillar tractor fitted with a massive 11-foot wide, 2.5-ton "Rome plow" blade equipped with a special 3-foot splitting lance or "stinger," and with 14 tons of added armor. A very limited number of the even more immense D-9 tractors are also in use. More than twice the size and weight of the D-7, each of these machines is said to be the operational equal of several. The tractors are presently organized into five companies of three platoons each, each company operating 30 or more tractors. Unofficially, the companies go under such names as Rome Runners, Land Barons, and Jungle Eaters. These outfits bulldoze continuously from dawn to dusk, seven days a week under what can only be described as spine-twisting and gut-wrenching (to say nothing of dangerous) conditions. No tree appears to be too large and no jungle too dense to escape these powerful machines in what must certainly be the most intense land-clearing program known to history.

The bulldozing began on a very small scale in 1965 and was devoted primarily to the clearing of roadsides and other lines of communication in order to discourage enemy ambushes. It was not until mid-1967 that the tractors were organized into small units. By the beginning of 1968, most of the major road systems in the central half of South Vietnam (Military Regions II and III) had already been cleared. Although this mission still continues, virtually all major roads in the country have now been cleared for 300 to 600 feet or more on each side. These swaths throughout forest and plantation are now a conspicuous feature of the Vietnamese landscape. In some instances chemical herbicide treatment has helped to maintain these strips in a treeless condition.

The employment of massed tractors organized into companies for extensive forest clearing began in 1968, and the program has expanded ever since. In its primary mission of denying forest cover and sanctuary, the "Rome plow" appears to be without equal. Effectiveness of the tractors is clearly superior to that of aerial application of chemical antiplant agents. The devices are considered, for example, to be playing an instrumental role in the attempt to "secure" the region centered around Saigon (Military Region III). They are also of considerable importance in the northern half of the country (Military Regions I and II). The United States has outfitted and is training two Vietnamese land-clearing companies as one of the facets of "Vietnamization."

We were able to spend one day in action with the 984th Landclearing Company, which at the time was operating in the southeastern corner of Tay Ninh province. During our stay, the company was in the final stages of obliterating the Boi Loi woods. More accurately, it was supply the *coup de grace* to this longtime enemy stronghold that previously had been treated at least once with herbicides, had been subjected to saturation bombing from B-52 stratofortresses, and had also been shelled by artillery.

We joined the outfit on its twenty-seventh day in the Boi Loi woods. During
the past 26 it had already scraped clean 6,037 acres. Several days more and this
job would be finished, permitting the 984th to move on to greener pastures.
Before this job, the men had eliminated the 9,000-acre Ho Bo woods in nearby
west-central Binh Duong province.

The Boi Loi woods was enemy territory and we were dropped in by helicopter.
We accompanied one of the platoon commanders in his armored personnel
carrier and were flanked by several Sheridan tanks of the Eleventh Armored
Cavalry. Although we had no contact with the enemy that day and hit no landmines, we were informed that both were regular occurrences. In the past 26 days,
for example, several enemy attacks had been repulsed and the tractors had set
off no less than 37 mines in the course of their work. (Seven casualties from
landmines had been sustained during this period.)

In operation, the tractors were strung out in a long staggered formation,
the lead tractor being directed for much of the time by the company commander
circling overhead in a small helicopter. The large number of bomb craters made
the job of maneuvering the large tracked vehicles most difficult. The heat was
oppressive (hovering around 130 degrees F. in the tractor cabs) and the work was
truly arduous. But the morale of the men seemed very high, despite their
fifteen-hour work days, seven days a week, wet season and dry. The company
was proud of its abilities and accomplishments and, we are told, was among the
rare units in Vietnam without a drug problem.

At the time of our visit, the unbulldozed terrain was covered largely by a tangle
of head-high, broad-leafed brushy plants and vines intermingled with *Imperata*
grass and shrubby bamboos. Of the scattered trees, more than half were dead. The
plow blades were set to skim the surface, each tractor scraping bare almost an
acre per hour. The big trees came crashing to the ground with great regularity.
Most were simply pushed over, but the really large ones were first split by the
stinger.

The terrain was flat and the soil a heavy grey podzolic, so that neither erosion
nor laterization (hardening of soils to a brick-like substance) are likely to be
problems here. In view of available seed (or other reproductive plant parts) and
shadeless conditions, this area is likely to be quickly dominated by a combination
of *Imperata* grass and shrubby bamboos, thereby largely precluding reforestation
for years (perhaps decades) to come. On other areas we inspected in Binh Duong
province that had been bulldozed two or three years previously, by far the most
prevalent vegetation was the worthless and pernicious weed *Imperata*. Indeed,
of the thousands of acres of formerly bulldozed areas that we were able to see
on this and our previous visits, there was only one area where forest trees (a com-
mercially lowgrade species of *Dipterocarpus*) were recolonizing naturally. Where
bulldozing is done in more hilly terrain, erosion can become a severe liability.
Moreover, with the elimination of the enormous water-holding capacity of an
extant forest, the heavy rains characteristic of Vietnam can produce severe flood
damage. We learned of one devastating flash flood in a recently bulldozed area
in Khanh Hoa province.

It cannot be denied that there are advantages to the bulldozing, given the conditions of this grim war. First, bulldozing largely clears areas of landmines, an ever-present horror throughout much of Vietnam to all who attempt to reutilize a war-visited area. (One Vietnamese whom we came to know has so far lost six relatives to mines left behind by one side or the other.) Secondly, some of the timber can subsequently be salvaged, particularly for firewood and charcoal manufacture. Thirdly, some of the bulldozed lands in "secure" areas have been taken over for agricultural pursuits, although this is often not feasible even in such areas because of extensive craterization by explosives. Farming is particularly evident in the roadside strips near population centers. A small fraction of the clearing by bulldozers is actually said to be done with subsequent resettlement or agricultural pursuits in mind (see, for example, *New York Times,* July 15, 1971, p. 3).

Bulldozing has, according to official military sources, leveled over 750,000 acres to date. I estimate that clearing continues at a rate of more than 1,000 acres per day. Because I was unable to obtain a breakdown of land and land use categories that have fallen to the relentless bulldozers, it is difficult to estimate the overall economic loss that can be attributed to these operations. However, some partial indications can be presented. With respect to the timber resource, the South Vietnamese forest service has determined that at least 126,000 acres of prime timber lands accessible to lumber operations have been destroyed through 1970, together with an estimated twenty million board feet of marketable tropical hardwood timber. At recent Saigon market values, averaging about $72 per thousand board feet, this amounts to a loss of $14.7 million. To this sum must be added a future loss due to destruction of growing stock. With respect to the rubber resource, the French rubber interests in South Vietnam have determined that substantially more than 2,500 acres of producing rubber trees (representing just over 1 percent of South Vietnam's total rubber) have so far been destroyed by bulldozers. There are about 120 rubber trees per acre, with an average value of $88 per tree. Total loss here can thus be estimated to exceed $26.4 million. (I might add that to the consternation of the French owners, they have received no compensation.)

There are, of course, many other losses attributable to the bulldozer program, most of which are impossible to quantify. Among them can be listed site degradation, erosion, weed invasion, destruction of wildlife habitat, flood damage, and miscellaneous property loss. One recent press report from western Hau Nghia province tells of the obliteration by bulldozers of a still partially inhabited farming region and the consequent disruptive impact (*New York Times,* May 7, 1971, p. 5). Even whole villages have been obliterated (*Nation,* Oct. 23, 1967, p. 397). Discussions with professional Vietnamese foresters revealed yet another headache connected with the land-clearing operations. In its nationwide forest conservation program, the South Vietnamese forest service issues timber-cutting permits on a judiciously restricted basis. However, corrupt province chiefs have, during the past year or so, come to realize that a denied local cutting permit can often be circumvented by turning to the Vietnamese Defense Department and, for pretended reasons of military necessity, request that the area be designated for bulldozing. If the request is granted, the chiefs can then cut the timber for personal profit. Finally, another use of the bulldozers results in a small amount of additional and unnecessary damage. In their lighter moments the engineers occasionally turn to carving up the landscape for the sheer hell of it. Thus one can now find a U.S. First Infantry Division emblem, covering some 1,500 acres, carved into the landscape about 25 miles northwest of Saigon (*New York Times,* Apr. 5, 1970, p. 7); a giant peace symbol is similarly engraved near Hue (*Life* magazine, July 2, 1971, p. 72).

In conclusion, the question is raised—although not answered—of how much forest loss can be sustained by an area before the regional ecology is adversely affected to a substantial extent. Before the war, more than 25 million acres of South Vietnam were covered by forest, representing about 60 percent of the country's total area of 43 million acres. So far, the war has claimed at the very least 3 million acres of the forest cover. The herbicide program has accounted for somewhat more than a third of this, the bulldozing somewhat less than a third, and the bomb, rocket, and shell craters (plus damage from other munitions) the remainder. Although the estimated 12 percent reduction in forest cover may not have a dramatic influence on the overall ecology of South Vietnam, detailed investigation will elucidate the magnitude of the subtle changes that have resulted. In the numerous local areas of severe damage, often covering several thousands of contiguous acres, the repercussions—both ecological and sociological—will be profound. (*Arthur H. Westing*)

The Big Bomb

In this report I summarize our findings about a new bomb in the United States arsenal, a bomb unique to the Second Indochina War. Owing to the paucity of information domestically available, I describe in some detail the bomb's general characteristics and employment.

The BLU-82/B general-purpose high-explosive concussion bomb turns out to be one of the most awesome and least publicized weapons to have been spawned by the war. It is a bomb with record-breaking dimensions: It is 4.5 feet in diameter, over 11 feet long, and weighs 15,000 pounds. Within its thin steel case are 12,600 pounds of a special, dense blasting agent (DBA-22M) consisting of a gelled aqueous slurry of ammonium nitrate and aluminum powder (plus a binding agent). This formulation provides a concussive blast surpassed only by that of a nuclear bomb.

Often referred to in Vietnam as the "Daisy Cutter" and sometimes as the "Cheeseburger," this super bomb is delivered by C-130E aircraft (of the 463rd Wing of the Seventh Air Force flying out of Cam Ranh Bay air base). Even though the bomb is floated to the ground by parachute from altitudes of 7,000 to 10,000 feet and occasionally even over 20,000 feet, the Seventh Air Force claims that the point of impact seldom is more than 300 feet off target and usually less than 150 feet. Timing of the drop is determined by ground radar.

The Daisy Cutter was developed primarily for the instant creation of clearings in dense jungle. Such clearings can then be used immediately as a landing zone by assault helicopters in locations inaccessible to conventional land-clearing equipment and techniques. The progenitor of this unique bomb was the 10,000 pound "blockbuster" bomb of World War II. Several dozen or more of these bombs were left over from that conflict. These were used in Indochina on an experimental basis, apparently beginning in 1967. The presently used BLU-82/B was developed and became operational in early 1970.

The bomb is detonated by an impact fuse at the end of an attached three-foot probe which sets off the main charge simultaneously at both ends of the bomb just above the ground. If all goes well, the resulting radial blast leaves no crater, but rather uproots and blows away all trees and other obstructions—even in heavy jungle—to create a virtually perfect clearing about the size of a football field. Although the size of each clearing differs, of course, according to local conditions of terrain and vegetation, the average radius of the opening, according to the Air Force, is about 160 feet, and its area thus about two acres. No fires are reported to have been set by these bombs and only minimal charring occurs. The blast is spectacular: A mushroom cloud rises some 6,000 feet into the air, and light aircraft flying more than two miles from the explosion are badly shaken by the shockwave. The landing zone, suitable for landing within minutes of the blast, can accommodate one to several assault helicopters at a time. The military code name for such an operation is "Commando Vault."

According to the Seventh Air Force, the average rate of use of the Daisy Cutters in South Vietnam has been one to two per week in recent months. (We were made aware of five drops during a one-week period in mid-August.) Although the total number of drops to date is classified information, an official spokesman for the U.S. Military Assistance Command in Vietnam (MACV) informed me that it is well in excess of 100. One press report claims that 160 drops occurred prior to June of 1970 (*Los Angeles Times*, June 1, 1970, p. 20). Most of the drops in South Vietnam have occurred in the northern half of the country and in the delta region to the south (Military Regions I, II, and IV). Information on the drops in Cambodia and Laos was not made available to us.

In the briefing we received on the Commando Vault operations, an official spokesman for the Seventh Air Force stressed and restressed that use of the Daisy Cutters was restricted to the creation of landing zones and that they were nothing more than "explosive bulldozers." He went into some detail on how the local populace is always alerted prior to a drop. On the other hand, we learned from another local Air Force source that exceedingly strict security is always maintained before a drop to avoid alerting the enemy; the flight crew does not even receive the target location or drop time until just before takeoff.

The Daisy Cutter is officially designated as a general-purpose bomb and has been used in a number of ways in Indochina beyond the creation of landing zones. One Air Force report explains that the bomb can be used for road interdiction by triggering landslides. In the Hanoi press this past spring (in an otherwise unconfirmed report) there is a description of the obliteration of an entire hamlet in Laos by this means. We also learned from three independent sources (two military and one embassy) that the Daisy Cutter has been and is being employed against enemy or suspected enemy troop concentrations. Moreover, in one of the Commando Vault missions we inspected from the air, the bomb had been dropped, according to our official military guide, onto a suspected enemy rocket emplacement. This mission had been carried out in June of this year in Quang Nam province nineteen miles southwest of Da Nang. The antipersonnel use of this bomb has also been reported in the press (for example, *New York Times,* Apr. 13, 1971, p. 1; Apr. 15, 1971, p. 5; Apr. 18, 1971, p. E2). Press reports describe one additional use for the bomb, that of removing the thick jungle canopy above suspected enemy storage areas (*New York Times,* Apr. 15, 1971, p.5).

What is the environmental impact of a Daisy Cutter? Here I am chagrined to report that of the scores of Commando Vault missions, old and new, no site could be found by MACV in an area secure enough for us to visit. U.S. citizens are generally not aware that the National Liberation Front and its North Vietnamese allies control virtually all of the forest and other wild lands of South Vietnam. Moreover, we could find no one who had examined or even thought about these sites with ecology in mind, not even the MACV science advisor or his staff. Although the immediate overt impact is easy to surmise, the more subtle and long-term effects must await further study.

According to an official Seventh Air Force source, the blast of a Daisy Cutter is of such intensity that all terrestrial and arboreal wildlife (as well as any luckless humans) within a radius of approximately 3,280 feet are killed outright by the concussive shockwave. The lethal zone from one such bomb thus covers an area of about 776 acres. Beyond this circle of death, concussion injury diminishes to insignificance radially outward for a distance of another 1,640 feet or so. This larger area of both death and injury to wildlife thus encompasses about 1,746 acres per bomb. Assuming that the total number of bomb drops to date has been 150, the forest area totally eliminated by this means has been only about 300 acres. Of much greater concern, all the wildlife occupying 116,400 acres or more have been killed. The wildlife on again as much area have sustained injuries. The Daisy Cutter thus adds significantly to the already severe stress imposed by the war on Vietnam's wildlife.

With respect to the vegetation, my information on damage is less complete. The innermost circle of two acres is, of course, totally annihilated. (One press report claims that even the worms in the ground are killed in this zone [*Life* magazine, May 21, 1971, p. 41].) I suspect that damage to the flora beyond the central, cleared area becomes negligible within a modest distance, but actual extent of damage will require on-site investigation. Recolonization by plants in the central, cleared zone seems to be fairly rapid, at least in the delta region. An Air Force officer familiar with the delta told us that a Commando Vault landing zone blasted out in that region looks green from the air within several weeks; it often becomes unusable within several months because of the regrowth of brush. Past experience suggests that the upland clearings will be quickly invaded primarily by *Imperata* grass and/or a variety of low-growing, brushy bamboo species, all tenacious and worthless weeds.

The Daisy Cutter is, in the words of one military officer we met in Vietnam, "a super bomb with super punch." MACV has been using these bombs on a steady basis for more than a year and a half now, apparently with no mention of them in the official daily, weekly, or monthly war news summaries. One senior Seventh Air Force officer explained to us, "They have such a devastating effect that we hate to give them much publicity."

The Commando Vault 7.5 ton bombs provide just one more means by which we casually rearrange the environment of Indochina with little if any concern about either the immediate or the long-term impact on the ecology of the area. I am painfully aware of how little in the way of biological data this report contains, but in providing the first detailed account of this new, indiscriminately wide-area weapon for the open literature, I hope that it will stimulate the necessary wildlife and other ecological studies as conditions permit.
(*Arthur H. Westing*)

RENE DUBOS

So Human an Animal

René Dubos is Professor of Microbiology at The Rockefeller University in New York. Among his many books are *Man, Medicine and Environment* (1968), *Mirage of Health* (1959), and *Man Adapting* (1965). This selection is taken from *So Human an Animal* (1968). Suppose we agree that conservation is necessary and that with rationality and good will, we can find ways to accomplish it, what do we want to conserve? Dubos points out that we cannot return the earth to the state in which it existed before people lived on it; that effects of man on his environment need not be bad, and, indeed, are not always destructive. Tame the earth with love, he suggests.

I live in mid-Manhattan and, like most of my contemporaries, experience a love-hate relationship with technological civilization. The whole world is accessible to me, but the unobstructed view from my 26th-floor windows reveals only a confusion of concrete and steel bathed in a dirty light; smog is a euphemism for the mud that constantly befouls the sky and blots out its blueness. Night and day, the roar of the city provides an unstructured background for the shrieking world news endlessly transmitted by the radio.

Everything I eat, drink, and use comes from far away, or at least from an unknown somewhere. It has been treated chemically, controlled electronically, and handled by countless anonymous devices before reaching me. New York could not survive a week if accident or sabotage should interrupt the water supply during the summer or the electric current during midwinter. My life depends on a technology that I do not really understand, and on social forces that are beyond my control. While I am aware of the dangers this dependence implies, I accept them as a matter of expediency. I spend my days in the midst of noise, dirt, ugliness, and absurdity, in order to have easier access to well-equipped laboratories, libraries, museums, and to a few sophisticated colleagues whose material existence is as absurd as mine.

Our ancestors' lives were sustained by physical work and direct associations with human beings. We receive our livelihood in the form of anonymously computerized paper documents that we exchange for food, clothing, or gadgets. We have learned to enjoy stress instead of peace, excitement in lieu of rest, and to extract from the confusion of day-to-day life a small core of exhilarating experiences. I doubt that mankind can tolerate our absurd way of life much longer without losing what is best in humanness. Western man will either choose a new society or a new society will abolish him; this means in practice that we shall have to change our technological environment or it will change us.

The following remarks made during a discussion held at Massachusetts Institute of Technology bring out the problems posed by the adaptation of human values to technological development.

Harvey Cox: ". . . There are components of the situation which allow themselves to be addressed by technological answers. But I think there is this other one which I don't think the technological answers get to, and it has a little bit to do with a question about our basic philosophical assumptions about man, and what it means to be fulfilled."

Questions from the floor: "But our basic philosophical assumptions may be pretechnological in nature, and *one of the main problems of man today may be to readjust philosophical perspective to modern technology*" [italics mine–R.D.].

Adjusting man's philosophical perspective to modern technology seems to me at best a dangerous enterprise. In any case, the technological conditions under which we now live have evolved in a haphazard way and few persons if any really like them. So far, we have followed technologists wherever their techniques have taken them, on murderous highways or toward the moon, under the threat of nuclear bombs or of supersonic booms. But this does not mean that we shall continue forever on this mindless and suicidal course. At heart, we often wish we had the courage to drop out and recapture our real selves. The impulse to withdraw from a way of life we know to be inhuman is probably so widespread that it will become a dominant social force in the future.

To long for a human situation not subservient to the technological order is not a regressive or escapist attitude but rather one that requires a progressive outlook and heroic efforts. Since we now rarely experience anything directly and spontaneously, to achieve such a situation would require the courage to free ourselves from the constraints that prevent most of us from discovering or expressing our true nature.

Sensitive persons have always experienced a biological and emotional need for an harmonious accord with nature. "Sometimes as I drift idly along Walden Pond," Thoreau noted in his *Journal*, "I cease to live and begin to be." By this he meant that he then achieved identification with the New England landscape.

The passive identification with nature expressed by Thoreau's phrase is congenial to Oriental thought but almost antithetical to Western civilization. Oddly enough, Tagore, a Hindu, came much closer than Thoreau to a typical Occidental attitude when he wrote that the great love adventure of European civilization had been what he called the active wooing of the earth.

"I remember how in my youth, in the course of a railway journey across Europe from Brindisi to Calais, I watched with keen delight and wonder that continent flowing with richness under the age-long attention of her chivalrous lover, western humanity. . . .

"Robinson Crusoe's island comes to my mind when I think of an institution where the first great lesson in the perfect union of man and nature, not only through love but through active communication, may be learnt unobstructed. We have to keep in mind the fact that love and action are the only media through which perfect knowledge can be obtained."

The immense and continued success among adults as well as among children of *Le Petit Prince* by the French writer Antoine de Saint Exupéry (1900–1944) also reflects a widespread desire for intimate relationships with the rest of creation.

"On ne connait que les choses que l'on apprivoise, dit le renard. Les hommes n'ont plus le temps de rien connaître. Ils achêtent des choses toutes faites chez les marchands. Mais comme il n'existe point de marchands d'amis, les hommes n'ont plus d'amis. Si tu veux un ami, apprivoise-moi!"

In the popular English translation *The Little Prince*, this passage reads as follows:

" 'One only understands the things that one tames,' said the fox. 'Men have no more time to understand anything. They buy things already made at the shops. But there is no shop anywhere where one can buy friendship, and so men have no friends any more. If you want a friend, tame me.' "

The French verb *apprivoiser* as used by Saint Exupéry is not adequately rendered by "tame." *Apprivoiser* implies here, not mastery of one participant over the other, but rather a shared experience of understanding and appreciation.

Poetical statements do not suffice to create conditions in which man no longer feels alienated from nature and from other men. But they are important nevertheless, because literary expressions often precede or at least sharpen social awareness. Poets, novelists, and artists commonly anticipate what is to be achieved one or two generations later by technological and social means. The poet is the conscience of humanity and at his best he carries high the torch illuminating the way to a more significant life.

Tagore wrote of man's active wooing of the earth, and stated that "love and action are the only media through which perfect knowledge can be obtained." Saint Exupéry urged that we can know and enjoy only that which we tame through love. Both have thus propounded a philosophical basis for conservation policies.

From a sense of guilt at seeing man-made ugliness, and also for reasons that must reach deep into man's origins, most people believe that Nature should be preserved. The exact meaning of this belief, however, has not been defined. There is much knowhow concerning conservation practices but little understanding of what should be conserved and why.

Conservation certainly implies a balance among multiple components of Nature. This is a doctrine difficult to reconcile with Western civilization, built as it is on the Faustian concept that man should recognize no limit to his power. Faustian man finds satisfaction in the mastery of the external world and in the endless pursuit of the unattainable. No chance for a stable equilibrium here.

To be compatible with the spirit of Western culture, conservation cannot be exclusively or even primarily concerned with saving man-made artifacts or parts of the natural world for the sake of preserving isolated specimens of beauty here and there. Its goal should be the maintenance of conditions under which man can develop his highest potentialities. Balance involves man's relating to his total environment. Conservation therefore implies a creative interplay between man and animals, plants, and other aspects of Nature, as well as between man and his fellows. The total environment, including the remains of the past, acquires human significance only when harmoniously incorporated into the elements of man's life.

The confusion over the meaning of the word Nature compounds the difficulty of formulating a philosophical basis for conservation. If we mean the world as it would exist in the absence of man, then very little of it survives. Not even the strictest conservation policies would restore the primeval environment, nor would this be necessarily desirable or even meaningful if it could be done.

Nature is never static. Men alter it continuously and so do animals. In fact, men have long recognized that they play a creative role in shaping Nature. In his *Concerning the Nature of the Gods,* written during the last century of the pre-Christian era, Cicero boasted: "We are absolute masters of what the earth produces. We enjoy the mountains and the plains. The rivers are ours. We sow the seed and plant the trees. We fertilize the earth. . . . We stop, direct, and turn the rivers; in short, by our hands we endeavor by our various operations in this world to make it as it were another Nature."

For animals as well as for men, the kind of environment which is most satisfactory is one that they have shaped to fit their needs. More exactly, the ideal conditions imply a complementary cybernetic relationship between a particular environment and a particular living thing. From man's point of view, civilized Nature should be regarded not as an object to be preserved unchanged, not as one to be dominated and exploited, but rather as a kind of garden to be developed according to its own potentialities, in which human beings become what they want to be according to their own genius. Ideally, man and Nature should be joined in a nonrepressive and creative functioning order.

Nature can be tamed without being destroyed. Unfortunately, taming has come to imply subjugating animals and Nature to such an extent as to render them spiritless. Men tamed in this manner lose their real essence in the process of taming. Taming demands the establishment of a relationship that does not deprive the tamed organism—man, animal, or nature—of the individuality that is the *sine qua non* of survival. When used in the sense of the French *apprivoiser*, taming is compatible with the spirit of conservation.

There are two kinds of satisfactory landscape. One is Nature undisturbed by human intervention. We shall have less and less of this as the world population increases. We must make a strenuous effort to preserve what we can of primeval

Nature, lest we lose the opportunity to re-establish contact now and then with our biological origins. A sense of continuity with the past and with the rest of creation is a form of religious experience essential to sanity.

The other kind of satisfactory landscape is one created by human toil, in which, through progressive adjustments based on feeling and thought, as well as on trial and error, man has achieved a kind of harmony between himself and natural forces. What we long for is rarely Nature in the raw; more often it is a landscape suited to human limitations and shaped by the efforts and aspirations that have created civilized life. The charm of New England or of the Pennsylvania Dutch countryside is not a product of chance, nor did it result from man's "conquest" of nature. Rather it is the expression of a subtle process through which the natural environment was humanized in accordance with its own individual genius. This constitutes the wooing or the taming of nature as defined by Tagore and Saint Exupéry.

Among people of Western civilization, the English are commonly regarded as having a highly developed appreciation of Nature. But in fact, the English landscape at its best is so polished and humanized that it might be regarded as a vast ornamental farm or park. River banks and roadsides are trimmed and grass-verged; trees do not obscure the view but seem to be within the horizon; fore-grounds contrast with middle distances and backgrounds. The parklands with their clumps of trees on shaven lawns, their streams and stretches of ornamental waters achieve a formula of scenery designed for visual pleasure in the spirit of the natural conditions.

The highland zone of western Britain constitutes a vast and remote area, not yet occupied by factories and settlements, offering open space for enjoyment and relative solitude. Conservation groups are struggling to protect its moors not only from industry and farming but also from reforestation. Yet the moors which are now almost treeless were once covered with an abundant growth of forest. The replacement of trees by heath and moor was not a "natural" event but one caused by the continued activities of man and his domesticated animals. Deforestation probably began as far back as the Bronze Age; the process was accelerated during the Middle Ages by the Cistercian monks and their flocks of sheep; then the exploitation of mines took a large toll of trees needed for smelting fires. In brief, the pristine ecological systems of the oak forest that once covered the highlands were eliminated by human action, leaving as relics only a few herd of deer. For nineteenth- and twentieth-century man, highland nature means sheep-walk, the hill peat bog, and the grouse moor. But this landscape is not necessarily the natural and right landscape, only a familiar one.

Public attitude toward the moors is now conditioned by literary associations. This type of landscape, which exists in other parts of England, evokes *Wuthering Heights* and the Brontë sisters. Since the wild moors are identified with passionate and romantic human traits, to reforest the highlands seems to show disrespect for an essential element of English literary tradition. Similarly, the garrets of Paris, sordid as they are physically, are associated with bohemian life, Mimi in *La Bohème*, and the tunes and romance of *Sous les Toits de Paris*. Art and literature have become significant factors in the landscape ecology of the civilized world.

The effects of history on nature are as deeply formative and as lasting as are those of early influences on individual persons and human societies. Much of what we regard today as the natural environment in England was in reality modified by the school of landscape painting in the seventeenth century. Under the guidance of landscape architects, a literary and artistic formula of naturalism transformed many of the great estates and then brought about secondarily similar modifications in large sectors of the English countryside and even of the cities. The effects of esthetic perceptions that first existed in the minds of the seventeenth-century painters thus became incorporated into the English landscape and will certainly long persist, irrespective of social changes. Less fortunately, the future development of American cities is bound to be oriented and constrained by the gridiron pattern and the network of highways which have shaped their early growth.

Profound transformations of nature by human activities have occurred during historical times over most of the world. Such changes are not all necessarily desirable, but the criteria of desirability are poorly defined. Since nature as it exists now is largely a creation of man, and in turn shapes him and his societies, its quality must be evaluated in terms not of primeval wilderness but of its relation to civilized life.

In his illuminating book *The Machine in the Garden,* Leo Marx has richly illustrated the contradictory attitudes toward Nature that have characterized American culture from its very beginning. The eighteenth-century Europeans saw America as a kind of utopian garden in which they could vicariously place their dreams of abundance, leisure, freedom, and harmony of existence. In contrast, most nineteenth-century immigrants regarded the forests, the plains, and the mountains as a hideous wilderness to be conquered by the exercise of power and harnessed for the creation of material wealth.

For most people all over the world today, the American landscape still has a grandeur and an ugliness uniquely its own. Above and beyond their geologic interest and intrinsic beauty, the Rocky Mountains and the Grand Canyon of the Colorado, for example, have acquired in world consciousness a cultural significance even greater than that of the highland moors or of the Mediterranean Riviera. The beauty of America is in those parts of the land that have not yet been spoiled because they have not been found useful for economic exploitation. The ugliness of America is in practically all its urban and industrial areas.

The American landscape thus means today either the vast romantic and unspoiled wilderness, or billboards and neon signs among dump heaps. Urban and industrial ugliness is the price that America and other technological societies seem to be willing to pay for the creation of material wealth. From the wilderness to the dump appears at present to symbolize the course of technological civilization. But this need not be so, or at least we must act in the faith that technological civilization does not necessarily imply the raping of nature. Just as the primeval European wilderness progressively evolved into a humanized creation through the continuous wooing of the earth by peasants, monks, and princes, so we must hope that the present technological wilderness will be converted into a new kind of urbanized and industrialized nature worthy of being called civilized. Our material wealth will not be worth having if we do not learn to integrate the machine, the city, and the garden.

The English archaeologist Jacquetta Hawkes in *A Land* has surveyed the interplay between the people and landscape of Britain in the course of history. She presents the appealing thesis that some two hundred years ago England had come close to achieving a harmonious equilibrium between local industrial activities, the towns and villages, the farming country, and the wilderness.

"Recalling in tranquillity the slow possession of Britain by its people, I cannot resist the conclusion that the relationship reached its greatest intimacy, its most sensitive pitch, about two hundred years ago. By the middle of the eighteenth century, men had triumphed, the land was theirs, but had not yet been subjected and outraged. Wildness had been pushed back to the mountains, where now for the first time it could safely be admired. Communications were good enough to bind the country in a unity lacking since it was a Roman province, but were not yet so easy as to have destroyed locality and the natural freedom of the individual that remoteness freely gives. Rich men and poor men knew how to use the stuff of their countryside to raise comely buildings and to group them with instinctive grace. Town and country having grown up together to serve one another's needs now enjoyed a moment of balance."

Even if this picture of eighteenth-century England does not exactly fit historical reality, it expresses ideals that could serve as goals for technological civilization. First it illustrates that conquest, or mastery of the environment, is not the only approach to planning nor is it the best. Man should instead try to collaborate with natural forces. He should insert himself into the environment in such a manner that he and his activities form an organic whole with Nature.

Jacquetta Hawkes also reminds us that both the humanized landscape and the wilderness have a place in human life, because they satisfy two different but equally important needs of man's nature. Modern man retains from this evolutionary past some longing for the wilderness, even though civilization has given him a taste for farmland, parks, and gardens. Conservation policies must involve much more than providing amusement grounds for sightseers and weekend campers; they must be concerned with the biological and cultural aspects of the human past.

The huge urban areas of the modern world present problems far more complex than those of eighteenth-century England, because they are now the cradle and home of the largest percentage of mankind. We must therefore learn to provide in them means to satisfy the physiological and psychological needs of man, including those acquired during prehistory. No social philosophy of urbanization can be successful if it fails to take into account the fact that urban man is part of the highly integrated web that unites all forms of life. There have been many large cities in the past, but until recent times their inhabitants were able to maintain fairly frequent direct contacts with the countryside or with the sea.

Historical experience, especially during the nineteenth century, shows that urban populations are apt to develop ugly tempers when completely deprived of such contacts. In our own times race riots provide further illustrations of this danger. Saving Nature in both its wild and humanized aspects is an essential part of urban planning.

While the problems of urbanization appear immensely complex, there is reason for optimism in the fact that some of the most crowded urban areas are also among the healthiest and most peaceful abodes of mankind. In several immense cities, furthermore, a high level of civilization has been maintained for over a thousand years. This is not the result of an accident. The medieval cities were carefully planned from their inception and life in them was rigidly controlled. The historical development of large and small European cities shows that planning and control are compatible with organic urban growth.

Unfortunately, the words "planning" and "control" are in bad repute in the United States. At a recent conference on Man's Role in Changing the Face of the Earth one of the participants flatly stated that he regarded planners as akin to missionaries and preferred a world in which "there are a number of ways of living and loving and eating and drinking and building and planting and playing and singing and worshiping and thinking." Freedom of behavior is important of course not only for its own sake but also because it is a condition of continued social and individual growth. On the other hand, social life is impossible without limitations to freedom. Furthermore, creativity must always be expressed within certain restraints. Almost everything that we hold dear implies restraints—from the form of a sonnet to the design of an early New England town, from the preservation of ancient monuments to familial and marital relations.

The wooing of the land in the farming areas of northern Europe, in the Mediterranean hill towns, or in the Pennsylvania Dutch country was achieved through man's willingness to accept topography and climate as a guide to planning and behavior. Many American cities—Los Angeles is only one of them— are the largest centers of nonrestraint in the world, and most of their problems derive precisely from a misapplied interpretation of freedom. In urban planning, as in all aspects of life, we must learn to discover and accept the restraints inherent in man's nature and in the conditions of our times. Civilizations emerge from man's creative efforts to take advantage of the limitations imposed on his freedom by his own nature and by the character of the land. . . .

Recommended Readings

American Chemical Society, *Cleaning Our Environment: The Chemical Basis for Action,* Washington, D.C., 1969.

Ehrenfeld, E.W., *Biological Conservation* (New York: Holt, Rinehart and Winston, 1970).

Goldman, M.I., *Controlling Pollution: The Economics of a Cleaner America* (Englewood Cliffs, New Jersey: Prentice-Hall, 1967).

Fisher, J., Simon, N., and Vincent, J., *Wildlife in Danger* (New York: Viking, 1969).

Kormondy, E.J., *Concepts of Ecology* (Englewood Cliffs, New Jersey: Prentice-Hall, 1969).

Nash, R., *The American Environment* (Reading, Massachusetts: Addison-Wesley, 1968).

Rose, S., *CBW, Chemical and Biological Warfare* (Boston: Beacon Press, 1969).

Smith, F.E., *The Politics of Conservation* (New York: Pantheon House, 1966).

Watt, K.E.F., *Ecology and Resource Management* (New York: McGraw-Hill, 1968).

Whyte, W.H., *The Last Landscape* (New York: Doubleday, 1968).